Energy, Climate and the Environment Series

Series Editor: **David Elliott**, Emeritus Professor of Technology, Open University, UK

Titles include:

Luca Anceschi and Jonathan Symons (*editors*)
ENERGY SECURITY IN THE ERA OF CLIMATE CHANGE
The Asia-Pacific Experience

Philip Andrews-Speed
THE GOVERNANCE OF ENERGY IN CHINA
Implications for Future Sustainability

Gawdat Bahgat
ALTERNATIVE ENERGY IN THE MIDDLE EAST

Ian Bailey and Hugh Compston (*editors*)
FEELING THE HEAT
The Politics of Climate Policy in Rapidly Industrializing Countries

Mehmet Efe Biresselioglu
EUROPEAN ENERGY SECURITY
Turkey's Future Role and Impact

David Elliott (*editor*)
NUCLEAR OR NOT?
Does Nuclear Power Have a Place in a Sustainable Future?

David Elliott (*editor*)
SUSTAINABLE ENERGY
Opportunities and Limitations

Horace Herring and Steve Sorrell (*editors*)
ENERGY EFFICIENCY AND SUSTAINABLE CONSUMPTION
The Rebound Effect

Horace Herring (*editor*)
LIVING IN A LOW-CARBON SOCIETY IN 2050

Matti Kojo and Tapio Litmanen (*editors*)
THE RENEWAL OF NUCLEAR POWER IN FINLAND

Antonio Marquina (*editor*)
GLOBAL WARMING AND CLIMATE CHANGE
Prospects and Policies in Asia and Europe

Catherine Mitchell, Jim Watson and Jessica Whiting (*editors*)
NEW CHALLENGES IN ENERGY SECURITY
The UK in a Multipolar World

Catherine Mitchell
THE POLITICAL ECONOMY OF SUSTAINABLE ENERGY

Ivan Scrase and Gordon MacKerron (*editors*)
ENERGY FOR THE FUTURE
A New Agenda

Gill Seyfang
SUSTAINABLE CONSUMPTION, COMMUNITY ACTION AND THE NEW ECONOMICS
Seeds of Change

Benjamin K. Sovacool
ENERGY & ETHICS
Justice and the Global Energy Challenge

Joseph Szarka
WIND POWER IN EUROPE
Politics, Business and Society

Joseph Szarka, Richard Cowell, Geraint Ellis, Peter A. Strachan and Charles Warren (*editors*)
LEARNING FROM WIND POWER
Governance, Societal and Policy Perspectives on Sustainable Energy

David Toke
ECOLOGICAL MODERNISATION AND RENEWABLE ENERGY

Xu Yi-chong (*editor*)
NUCLEAR ENERGY DEVELOPMENT IN ASIA
Problems and Prospects

Xu Yi-chong
THE POLITICS OF NUCLEAR ENERGY IN CHINA

Energy, Climate and the Environment
Series Standing Order ISBN 978–0–230–00800–7 (hardback)
978–0–230–22150–5 (paperback)

You can receive future titles in this series as they are published by placing a standing order. Please contact your bookseller or, in case of difficulty, write to us at the address below with your name and address, the title of the series and one of the ISBNs quoted above.

Customer Services Department, Macmillan Distribution Ltd, Houndmills, Basingstoke, Hampshire RG21 6XS, England

Alternative Energy in the Middle East

Gawdat Bahgat
Professor, Near East South Asia, Center for Strategic Studies, National Defense University, USA

palgrave
macmillan

First published 2013 by
PALGRAVE MACMILLAN

Palgrave Macmillan in the UK is an imprint of Macmillan Publishers Limited, registered in England, company number 785998, of Houndmills, Basingstoke, Hampshire RG21 6XS.

Palgrave Macmillan in the US is a division of St Martin's Press LLC, 175 Fifth Avenue, New York, NY 10010.

Palgrave Macmillan is the global academic imprint of the above companies and has companies and representatives throughout the world.

Palgrave® and Macmillan® are registered trademarks in the United States, the United Kingdom, Europe and other countries

ISBN: 978–1–137–26457–2

This book is printed on paper suitable for recycling and made from fully managed and sustained forest sources. Logging, pulping and manufacturing processes are expected to conform to the environmental regulations of the country of origin.

A catalogue record for this book is available from the British Library.

A catalog record for this book is available from the Library of Congress.

Contents

Series Editor's Preface

Concerns about the potential environmental, social and economic impacts of climate change have led to a major international debate over what could and should be done to reduce emissions of greenhouse gases. There is still a scientific debate over the likely *scale* of climate change, and the complex interactions between human activities and climate systems, but, global average temperatures have risen and the cause is almost certainly the observed build up of atmospheric greenhouse gases.

Whatever we now do, there will have to be a lot of social and economic adaptation to climate change – preparing for increased flooding and other climate-related problems. However, the more fundamental response is to try to reduce or avoid the human activities that are causing climate change. That means, primarily, trying to reduce or eliminate emission of greenhouse gasses from the combustion of fossil fuels. Given that around 80% of the energy used in the world at present comes from these sources, this will be a major technological, economic and political undertaking. It will involve reducing demand for energy (via lifestyle choice changes – and policies enabling such choices to be made), producing and using whatever energy we still need more efficiently (getting more from less), and supplying the reduced amount of energy from non-fossil sources (basically switching over to renewables and/or nuclear power).

Each of these options opens up a range of social, economic and environmental issues. Industrial society and modern consumer cultures have been based on the ever-expanding use of fossil fuels, so the changes required will inevitably be challenging. Perhaps equally inevitable are disagreements and conflicts over the merits and demerits of the various options and in relation to strategies and policies for pursuing them. These conflicts and associated debates sometimes concern technical issues, but there are usually also underlying political and ideological commitments and agendas which shape, or at least colour, the ostensibly technical debates. In particular, at times, technical assertions can be used to buttress specific policy frameworks in ways which subsequently prove to be flawed.

The aim of this series is to provide texts which lay out the technical, environmental and political issues relating to the various proposed policies for responding to climate change. The focus is not primarily on the science of climate change, or on the technological detail, although there will be accounts of the state of the art, to aid assessment of the viability of the various options. However, the main focus is the policy conflicts over which strategy to pursue.

The present volume looks at an area of the world where energy-related conflicts seem endemic and crucial in geo-political terms – the Middle East. It is also an area where, in some cases, rapid political change is underway. That in turn could lead to new pathways being explored in relation to energy. While oil and gas will continue to dominate, the region has a very large renewable energy resource, which is now being explored, and some countries within the region are also looking to nuclear power as a new energy option. This book looks at motivation for these changes. They include a growing awareness that demand for energy will rise as the economies expand and climate change impacts, with for example the energy needed for air conditioning and desalination beginning to divert significant proportion of the oil that would otherwise be exported. In addition, the wider global climate implications of continued reliance on fossil fuels suggest that diversification would be wise, with the possibility emerging that the region could become a major exporter of solar derived energy, whether in the form of electricity or synfuels made using renewable power sources.

In a region blessed with a very large and reliable solar input, it might seem odd that there is also an interest in nuclear power, but some see this as more reliable and as having strategic significance. This book summarises the state of play by reviewing the overall options, and then by exploring policy developments and debates in key selected countries. Clearly there is no shortage of potential for policy debate, even leaving out the political, economic and resource-base differences between and amongst the countries in this region.

Preface and Acknowledgements

Energy is the lifeblood of civilization. As both individuals and nation-states we depend heavily on energy. In almost everything we do, we rely on one or several sources of energy. Many people and governments used to take the availability of energy sources for granted. Our deepening reliance on energy and the rise of a combination of geopolitical, geological and environmental challenges have cast doubt on this assumption that energy will always be there. Little wonder that energy security has become a major concern to almost all countries in the world.

I have been working on energy for more than two decades. The policy of energy, at national and international levels, and the growing literature are immensely stimulating. Like many researchers, most of my work has focused on fossil fuels, particularly oil and natural gas. However, these fossil fuels are finite. The argument whether oil production has peaked is, in my opinion, irrelevant. What is certain is that these fossil fuels will not last forever. Furthermore, they do contribute to polluting the environment. Against this background, alternative energy, particularly renewable and nuclear power, has received substantial attention in recent years.

The Middle East holds massive proven reserves of oil and natural gas and for decades the region has been a major producer and exporter. In recent years several Middle Eastern states have sought to reduce their heavy dependency on these fossil fuels and diversify their energy mix. Strategic and economic forces are behind this strategy of diversifying the energy mix. Generally, it will take some time and a lot of money to develop the necessary human and technological infrastructure to support nuclear power. On the other hand, utilizing renewable resources such as hydro, solar and winds require less financial assets. Indeed, according to a report published by the International Renewable Energy Agency (IRENA) in November 2012, renewable power generation technologies are increasingly cost-competitive. Generally, improved technology and mass production tend to push prices down. On the positive side, renewable resources have tremendous potential in most Middle Eastern countries.

In the introductory chapter, I discuss the main characteristics of nuclear and renewable energy. The next six chapters examine the energy outlook in six case studies: Morocco, Egypt, Israel, Saudi Arabia, the

United Arab Emirates and Iran. The focus is on nuclear and renewable energy with references to oil and natural gas sectors. Given space and time limitations, these six countries give good representation of energy policy in the Middle East. Politically and economically they differ from each other. Morocco, Egypt and Israel hold smaller hydrocarbon deposits than Saudi Arabia, the United Arab Emirates and Iran. Still, all of them have recently adopted strategies to diversify their energy mixes. As the following chapters demonstrate, some of these countries (i.e. Iran and the UAE) have been more aggressive and more successful than the others. Together these six case studies highlight the similarities and diversities in Middle Eastern efforts to utilize alternative energy.

One of the main arguments in this volume is that each source of energy has its own advantages and disadvantages. The argument is not that one source should replace another; rather, it is that diversifying the energy mix would address economic, strategic and environmental concerns. Another major focus is on conservation. The energy equation has two sides: supply and demand. Diversifying energy sources is crucial. Efforts should be made simultaneously to curb consumption. Energy consumption in the Middle East is rising at an alarming rate. Many countries have already implemented policies to address this challenge.

Writing a book is a huge adventure, with so many ups and downs. Professional and personal support from family and close friends is crucial in this undertaking. I would like to thank Helen Hooker, Sandra Dickson, Beth Sims, Theresa McDevitt, Helen Wedlake and Patrizia Bassani. Finally, I would like to thank my friends and colleagues at the Near East South Asia Center for Strategic Studies, the National Defense University. Despite all the assistance I have received in the course of writing this book, all errors of facts or judgement are mine alone.

List of Abbreviations

ADCO	Abu Dhabi Company for Onshore Operations
ADMA-OPCO	Abu Dhabi Marine Operating Company
ANDATCO	Abu Dhabi National Tanker Company
ADNOC	Abu Dhabi National Oil Company
AEOI	Atomic Energy Organization of Iran
AGP	Arab Gas Pipeline
ARAMCO	Arabian-American Oil Company
BcF	Billion Cubic Feet
B/D	Barrel per Day
BG	British Gas
BIRD	Bi-national Industrial Research and Development
BOO	Build–Own–Operate
BP	British Petroleum
CDER	Renewable Energy Development Center
CNR	Compressed Natural Gas
CO_2	Carbon Dioxide
CSP	Concentrated Solar Power
DII	Desertec Industrial Initiative
ECHEM	Egyptian Petrochemicals Holding Company
EEZ	Exclusive Economic Zone
EEU	Energy Efficiency Unit
EGAS	Egyptian Natural Gas Holding Company
EGPC	Egyptian General Petroleum Corporation
EIA	Energy Information Administration
EMRA	Egyptian Mineral Resources Authority
ENEC	Emirates Nuclear Energy Corporation
EOR	Enhanced Oil Recovery
EU	European Union
EURODIF	European Gaseous Diffusion Uranium Enrichment Consortium
FANR	Federal Authority for Nuclear Regulation
GANOPE	Ganob El-Wadi Petroleum Holding Company
GASCO	Abu Dhabi Gas Industries
GCC	Gulf Cooperation Council
GCCIA	Gulf Cooperation Council Interconnection Authority
GDP	Gross Domestic Product

GECF	Gas Exporting Countries Forum
GHG	Green House Gas
GICNT	Global Initiative to Combat Nuclear Terrorism
GIS	Geographic Information System
GIZ	Gesellschaft für Internationale Zusammenarbeit
GNEII	Gulf Nuclear Energy Infrastructure Institute
HE	Hydrogen Energy
HEU	High-enriched Uranium
HLW	High-level Waste
HVDC	High Voltage Direct Current
IAB	International Advisory Board
IAEA	International Atomic Energy Agency
IAEC	Israel Atomic Energy Agency
IEA	Internationial Energy Agency
IGD	Integrated Gas Development
IHA	International Hydropower Association
IMF	International Monetary Fund
INPO	Institute of Nuclear Power Operations
IOCs	International Oil Companies
IRS	Incident Reporting System
JV	Joint Venture
KA-CARE	King Abdullah City for Atomic & Renewable Energy
KAUST	King Abdullah University of Science & Technology
KEPCO	Korea Electric Power Company
LEU	Low-enriched Uranium
LNG	Liquefied Natural Gas
LPG	Liquefied Petroleum Gas
MASEN	Moroccan Agency for Solar Energy
MEG	Meghreb-Europe Gas Pipeline
MENA	Middle East North Africa
MGS	Master Gas System
MIST	Masder Institute of Science and Technology
MIT	Massachusetts Institute of Technology
MMT	Million Metric Tons
MoN	Memorandum of Understanding
MSP	Mediterranean Solar Plan
MW	Megawatt
NAMAs	Nationally Appropriate Mitigation Actions
NEA	Nuclear Energy Agency
NGSCO	National Gas Shipping Company
NGV	Natural Gas Vehicles

NIGC	National Iranian Gas Company
NIOC	National Iranian Oil Company
NIORPDC	National Iranian Oil Refining Production & Distribution Company
NPP	Nuclear Power Plants
NPT	Non-Proliferation Treaty
NREA	New and Renewable Energy Agency
NSCSA	National Shipping Company of Saudi Arabia
NSS	National Solar System
OECD	Organisation for Economic Cooperation and Development
One	National Electricity Office
OPEC	Organization of Petroleum Exporting Countries
POGC	Pars Oil and Gas Company
PPP	Public–Private Partnership
PSA	Production-Sharing Agreement
PV	Photovoltaic
R & D	Research and Development
RE	Renewable Energy
REAI	Renewable Energy Association of Israel
SCA	Suez Canal Authority
SCIP	Seven Countries Interconnection Project
SPC	Supreme Petroleum Council
SUMED	Suez-Mediterranean Pipeline
SUNA	Renewable Energy Organization of Iran
SWP	Sahara Wind Project
TcF	Trillion Cubic Feet
TNRC	Tehran Nuclear Research Center
TOE	Ton Oil Equivalent
TREC	German Trans-Mediterranean Renewable Energy Cooperation
TSC	Technical Services Contract
UAE	United Arab Emirates
UfM	Union for the Mediterranean
ULCCs	Ultra Large Crude Carriers
UNDP	United Nations Development Programme
UNIFIL	United Nations Interim Force in Lebanon
UOG	United Arab Emirates Offsets Group
VLCCs	Very Large Crude Carriers
WAND	World Association of Nuclear Operators
WINS	World Institute for Nuclear Security

Maps

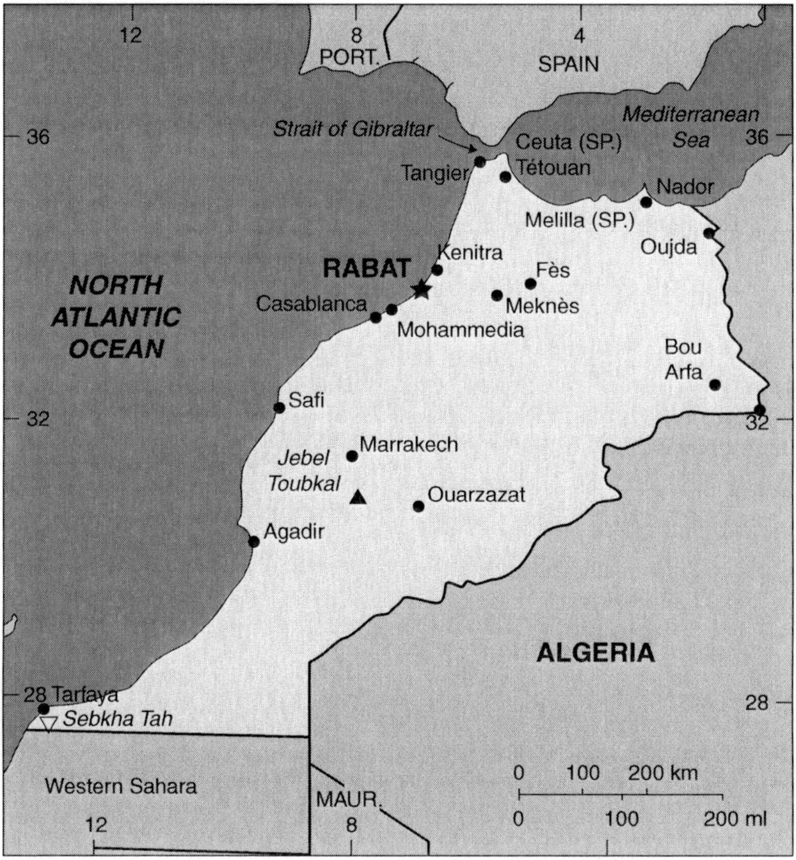

1 Morocco

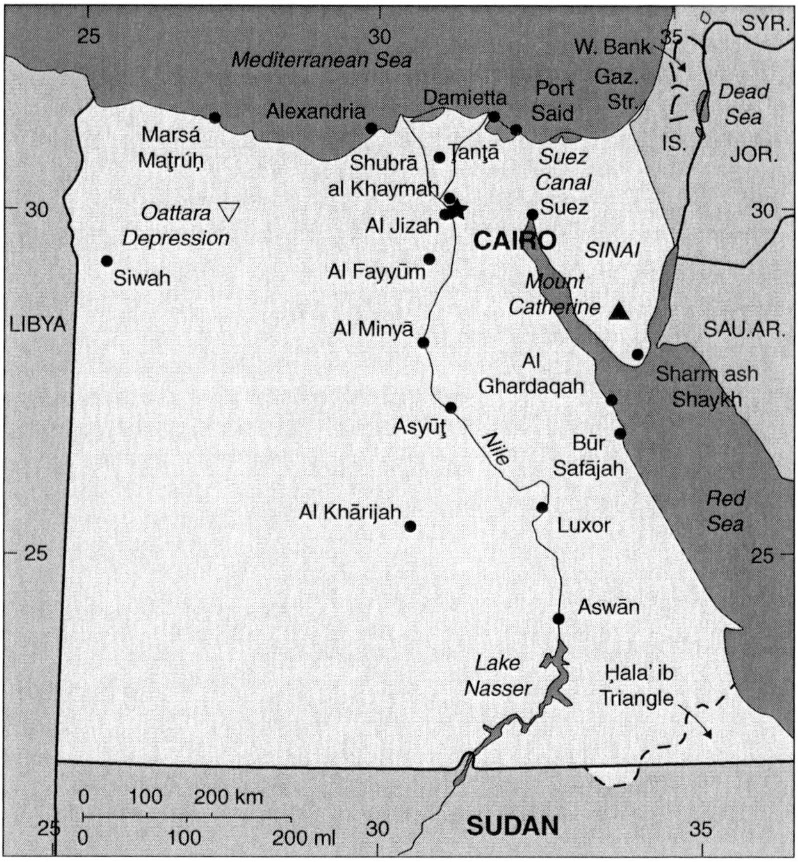
25
30
35
SYR.
Mediterranean Sea
W. Bank
Gaz.
Str.
Dead
Sea
Port
Said
Damietta
Alexandria
Marsá
Maţrúḩ
Ţanţā
Shubrā
al Khaymah
Suez
Canal
IS.
JOR.
30
Oattara
Depression
Suez
Al Jizah
CAIRO
SINAI
Siwah
Al Fayyūm
Mount
Catherine
LIBYA
Al Minyā
SAU.AR.
Al
Ghardaqah
Sharm ash
Shaykh
Asyūţ
Nile
Būr
Safājah
Red
Sea
Al Khārijah
Luxor
25
Aswān
Lake
Nasser
Ḩala' ib
Triangle
0 100 200 km
0 100 200 ml
SUDAN

2 Egypt

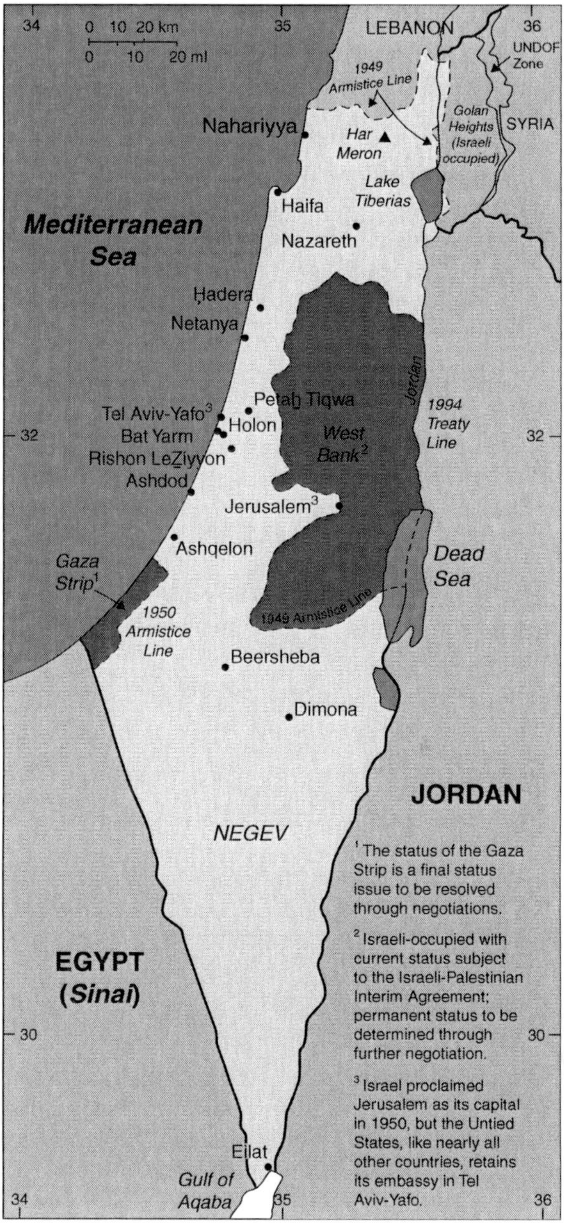

34
0 10 20 km
0 10 20 ml
35
LEBANON
36
UNDOF Zone
1949 Armistice Line
Nahariyya
Har Meron
Golan Heights (Israeli occupied)
SYRIA
Lake Tiberias
Haifa
Mediterranean Sea
Nazareth
Ḥadera
Netanya
Jordan
Petah Tiqwa
Tel Aviv-Yafo³
Holon
1994 Treaty Line
32
Bat Yam
West Bank²
Rishon LeZiyyon
Ashdod
Jerusalem³
Ashqelon
Dead Sea
Gaza Strip¹
1949 Armistice Line
1950 Armistice Line
Beersheba
Dimona
JORDAN
NEGEV
¹ The status of the Gaza Strip is a final status issue to be resolved through negotiations.
² Israeli-occupied with current status subject to the Israeli-Palestinian Interim Agreement; permanent status to be determined through further negotiation.
³ Israel proclaimed Jerusalem as its capital in 1950, but the Untied States, like nearly all other countries, retains its embassy in Tel Aviv-Yafo.
EGYPT (Sinai)
30
Eilat
Gulf of Aqaba

3 Israel

4 Saudi Arabia

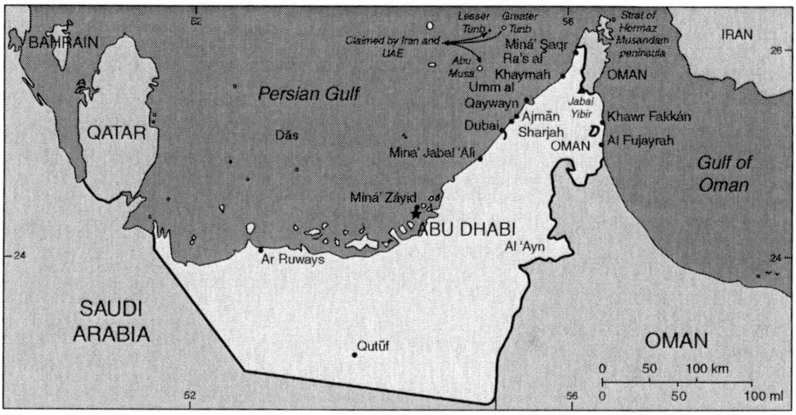

5 United Arab Emirates

UZB.
ARM.
AZER.
50
Caspian Sea
40
TURKMENISTAN
Tabriz
Lake Urmia
Now Shahr
Mashhad
TEHRAN
Kūh-e Damavand
Qom
Kermānshāh
Tigris
Birjand
AFG.
Dezfūl
Eşfahān
IRAQ
Yazd
Euphrates
Ahvàz
Ābàdàn
30
Kermān
30
Shiràz
KUW.
Zāhedān
Shatt al Arab
Bandar-e Búshehr
Bandar 'Abbās
'Asalúyeh
SAUDI ARABIA
Persian Gulf
Chāh Bāhar
PAK.
BAHR.
Strait of Hormuz
QATAR
OMAN
0 100 200 km
0 100 200 ml
50
Gulf of Oman
60
U.A.E.

6 Iran

1
Introduction

For most of the modern era fossil fuels – oil, natural gas, and coal – have provided the lion's share of global energy supplies. Their relative shares have fundamentally changed, but they have continued to dominate the energy mix in almost every country in the world. This trend is likely to prevail in the foreseeable future. However, other fuels, particularly nuclear power and renewable energy, such as that derived from wind and solar have attracted substantial attention and investments and are projected to provide an incremental share of supplies in the coming years and decades. According to a recent report by ExxonMobil oil, gas and coal will make up about 80 per cent of total energy consumption in 2040. Nuclear power will grow on average at about 2.2 per cent a year, a substantial increase, but lower than projections prior to the 2011 Fukushima disaster in Japan. Finally, wind, solar, and other renewable resources will see strong growth and account for about 4 per cent of global demand by 2040.[1]

More than any other region in the world, the broad Middle East is blessed with substantial hydrocarbon resources. Together, Middle Eastern states hold approximately 52 per cent of the world's total proven oil reserves and 42.4 per cent of proven gas reserves.[2] These huge reserves mean that the region is not likely to run out of oil or gas any time soon. This conclusion, however, does not guarantee the region's energy security or economic prosperity. In recent years, a combination of population explosion, high rate of economic growth, and heavy government energy subsidies has led to a surge in energy consumption all over the Middle East. Indeed, the region has one of the highest energy consumption rates per capita in the world.

This surge in energy consumption has negatively impacted the region both environmentally and economically. Several Middle Eastern cities are among the most polluted in the world (largely due to the burning of cheap oil). At the same time, rising consumption means less oil and gas available for export. Despite efforts to diversify their economies away from oil and create other sources of national income, most Middle Eastern countries are still heavily dependent on oil (and to a lesser extent gas) revenues. This means that higher consumption leads to less export and shrinking revenues. These shrinking revenues would open the door to socio-economic and political uncertainties and turmoil.

In addition to these environmental and economic consequences, Middle Eastern leaders understand that oil and gas are finite resources. Like the rest of the world, most Middle Eastern countries have joined the global search for alternative sources of energy. The two most promising sources are nuclear power and renewable resources. In recent years several Middle Eastern governments have taken significant steps to utilize energy derived from alternative sources. In the short term, the Middle East region, like the rest of the world, is likely to remain dependent on oil and gas. However, in the long term, the development of alternative sources of energy (e.g., nuclear and renewables) is likely to enhance the region's energy security, leave more oil and gas for export, and contribute to reducing pollution.

This chapter is divided into two parts. The first discusses the pros and cons of nuclear power and the second examines those of renewable sources. Like all other sources of energy, nuclear and renewables have their own advantages and disadvantages. In the coming years and decades oil and natural gas are likely to remain the more reliable sources of energy in the Middle East. The development of nuclear and renewables is likely, at least partly, to improve the region's collective energy security. The underlying conclusion of this study is that the diversification of the energy mix and the enhancement of energy efficiency are the most effective strategies.

Nuclear power

The birth of civilian nuclear energy was marked by the first nuclear electricity obtained in a US nuclear laboratory in 1951 and the first nuclear power plant created by academician Igor I. Kurchatov in the Soviet Union in 1954.[3] Since then nuclear power has been used in electricity generation as well as industrial and medical applications. The excitement and expansion of the first two decades of the nuclear era were replaced by virtual stagnation during most of the 1980s and 1990s. Since the dawn of the new millennium, however, the renewed interest in nuclear power has intensified in what has been dubbed the 'nuclear renaissance' – ambitious plans to build new nuclear plants or expand the life of existing ones. In 2010, 29 countries operated 441 plants, with a total capacity of 375 megawatts. These plants produced slightly less than 14 per cent of the world's electricity supplies and 5.7 per cent of total primary energy. A further 60 plants, with a generating capacity of 58.6 megawatts, are currently under construction.[4]

This global interest in nuclear power can be explained by at least four overlapping developments. First, population growth and economic development in many countries have fuelled the demand for energy. This surge in demand has not been sufficiently met due to the volatility of oil and gas prices and some interruptions of supplies (i.e. Russia's gas to Ukraine and Western Europe). The volatility of prices and uncertainty of supplies have underscored energy security vulnerability of several countries and the need to diversify their energy mix away from fossil fuels. Second, the growing concern about greenhouse gas (GHG) emissions, particularly carbon dioxide from the combustion of fossil fuels, has created renewed interest in the basically carbon-free nuclear power. In short, nuclear power is considered an important part of the global effort to address climate change. Third, since the dawn of the nuclear era, nuclear power has been utilized mainly in developed countries. It is little wonder that the initiation and development of nuclear programs have been seen in many under-developed and developing countries as symbols of technological advance. Finally, the improved performance of existing reactors has alleviated the concern some countries have over the security and reliability of nuclear power. In the last few decades nuclear power has been proven and seen as a non-intermittent and readily expandable source of energy.

These recent positive developments should not be overestimated. Nuclear power has continued to face daunting challenges and risks that need to be addressed. The list includes the high construction costs, safety, waste, and the close connection between civilian nuclear power and military applications. These largely unresolved hurdles have cast a doubt on the prospects of a nuclear renaissance. Thus, many countries have expressed interest in diversifying their energy mix away from fossil fuels but a few have articulated specific plans and concrete steps to overcome the serious uncertainties regarding nuclear energy.

Since the early 2000s a large number of Middle Eastern countries, and many others all over the world particularly in Asia, have expressed interest in establishing and developing nuclear programs. Like other regions, environmental, financial, and strategic concerns are driving nuclear interest in the Middle East. Due to heavy energy subsidies, many Middle Eastern cities are among the most polluted in the world. Replacing oil by nuclear power for generating electricity and water desalinization would release more oil for export and increase national revenues. Finally, Arabs' and Turks' interest in nuclear power can be seen as a potential hedging strategy against Iran's nuclear program. The progress Iran has achieved in advancing its own nuclear program has both galvanized its Arab and Turkish neighbours into launching their own programs and made it easier for them to justify doing so.[5] The relative weight of each drive varies from one Middle Eastern country to another. In addition, how they will address the uncertainties surrounding nuclear power is yet to be spelled out. What is certain is that the growing interest in nuclear power in the Middle East has the potential to drastically alter the security and strategic landscape in the entire region with significant implications worldwide.

Challenge # 1: nuclear safety

The full utilization of nuclear power, and indeed its future, depends, to a large extent, on public perception and public confidence in the safety of nuclear plants. The threats to safety come mainly from at least three sources: (1) unintended technical accidents caused by equipment failure or human errors, (2) natural disaster such as earthquake or tsunami, and/or (3) potential planned terrorist attacks. In order to address these challenges several initiatives at national, regional, and international levels have been taken. As a result, nuclear safety has substantially improved. However, the risk of an accident or terrorist attack is not zero. Furthermore, these measures to improve safety have contributed to the rising of plant construction and operation.

The most serious nuclear accidents due to mechanical failure are the Three Mile Island in the United States in 1979 and the Chernobyl disaster in the former Soviet Union in 1986. In the Three Mile Island nuclear plant a cooling malfunction caused part of the core to melt in the # 2 reactor. Some radioactive gas was released a couple of days after the accident, but not enough to cause any dose above background levels to local residents. There were no injuries or adverse health effects from the accident.[6] The Chernobyl disaster was the product of a flawed Soviet reactor design coupled with serious mistakes made by the plant operators. Acute radiation syndrome was diagnosed in 237 people on-site and those involved with the clean-up. A large proportion of childhood thyroid cancers diagnosed since the accident is likely to be due to intake of radioactive iodine fallout. Further, large areas of Belarus, Ukraine,

Russia, and beyond were contaminated to varying degrees.[7] These two accidents dealt a heavy blow to the expansion of nuclear power not only in the two respective nations but all over the world. In the ensuing years the construction of new plants came to a halt and some countries closed, or considered closing, their nuclear reactors.

The Tohoku earthquake and tsunami on 11 March 2011 in Japan resulted in a series of equipment failures, nuclear meltdowns, and releases of radioactive materials at the Fukushima Nuclear Power Plant. There were no immediate deaths due to direct radiation exposures. In December 2011 Japanese authorities declared the plant to be stable, though it would take decades to de-contaminate the surrounding areas. This disaster has raised doubt and uncertainty over the future of nuclear power. Prior to this tragedy, it appeared that the world had successfully overcome the 'Chernobyl syndrome'.[8] There has not been a universal response. Rather, some countries, such as Germany, have decided to suspend or phase out their nuclear reactors, while others (such as Russia, China, and India) have not changed their ambitious plans to expand their nuclear programs.

On the positive side, international efforts to promote cooperation and improve safety have intensified. For example, the Institute of Nuclear Power Operations (INPO) was established in 1979 as a not-for-profit organization headquartered in Georgia in the United States. The INPO conducts corporate evaluations that focus on safety and reliability. It also provides training and support for nuclear power professionals. Through information exchange and publications, the INPO communicates lessons learned and best practices throughout the nuclear power industry. Finally, at the request of individual nuclear-electricity-generating facilities, the INPO provides assistance with specific technical or management issues in areas related to plant operation and support.[9] Similarly, the London-based World Association of Nuclear Operators (WANO) was founded in 1989. Its mission is to maximize the safety and reliability of nuclear power plants worldwide by working with other countries to assess benchmark and improve performance through mutual support, exchange of information, and emulation of best practices.[10]

Equally important, the International Atomic Energy Agency (IAEA) has actively sought to create a universal, effective, and transparent network for sharing operating experience and promoting communication about design deficiencies, human errors, and near misses. The IAEA jointly with the Nuclear Energy Agency (NEA) operates an Incident Reporting System (IRS). Through IRS, information is collected from around the world on unusual events in nuclear power plants that may be important for safety or accident prevention and actions. The information is then assessed, analysed, and fed back to operators to prevent similar occurrences at other plants. The ultimate objective is reducing the frequency and severity of safety significant unusual events at nuclear power plants.[11]

A close examination of nuclear accidents, caused either by nature or human error, suggests that the few ones that had already occurred had significant impact on nuclear industry worldwide. Global public perception and confidence play a key role in launching and developing nuclear programs. The experience of the last several decades indicates that significant improvements have been made. As a result, nuclear plants are much safer today than they were a few decades ago. However, these safety improvements have added additional costs to nuclear plants. These safety improvements and additional costs aside, the risk of major nuclear accidents is not zero and should not be ruled out.

Terrorist threats to nuclear plants are multi-dimensional. A terrorist sabotage can cause a major radioactive release. Terrorists can seek to obtain nuclear material or device. Not surprisingly, the protection of nuclear installations has always been a major concern. The revelations in recent years that some terrorist organizations considered crashing a jumbo jet in a nuclear plant have heightened the alarm.

Several initiatives have been taken to counter these threats. Since the early 1990s the Nunn-Lugar Cooperative Threat Reduction Program and similar efforts have succeeded in securing nuclear material in dozens of sites in the former Soviet Union and elsewhere. In a landmark development in 2004, the United Nations Security Council unanimously approved Resolution 1540, which legally requires all states to refrain from supporting by any means non-state actors from developing, acquiring, manufacturing, possessing, transporting, transferring, or using nuclear, chemical, or biological weapons and their delivery systems. It also requires all states to provide 'appropriate effective' security for any stockpiles of nuclear weapons or related materials they may have.[12] More than 80 countries, including Bahrain, Jordan, Libya, Morocco, Saudi Arabia, Turkey, and the United Arab Emirates (UAE), are members in the Global Initiative to Combat Nuclear Terrorism (GICNT), which seeks to foster international cooperation in order to prevent terrorists from acquiring, transporting, or using nuclear materials and radioactive substances, or carrying out hostile actions against nuclear facilities.[13] Finally, in 2008 the World Institute for Nuclear Security (WINS) was established to bring together nuclear security experts, nuclear industry, governments, and international organizations to focus on rapid and sustainable improvement of security at nuclear facilities around the world.[14]

This brief discussion of nuclear safety suggests that even in countries with a long history of building and operating nuclear plants there is room for significant improvement. Countries with ambition to benefit from nuclear power, in the Middle East and elsewhere, have limited experience and largely lack the necessary human and technological infrastructure to construct and operate nuclear plants in a safe fashion. Furthermore, the rush to build power plants in conjunction with weak national regulatory systems and shortages of trained personnel may 'exacerbate concerns about safety in the future'.[15] In addition, more nuclear plants in many countries would mean more potential targets for terrorist sabotage and higher probability that a terrorist attack would succeed where a rigid and solid security is lacking.[16] These concerns indicate that the IAEA is likely to play a leading role in helping the newcomer states put in place the necessary infrastructure needed to develop nuclear power safely.

Challenge # 2: waste management

In addition to the challenge of how to secure the safety of nuclear plants, public concern over how to manage nuclear waste is a major constraint on large-scale expansion of nuclear power. Radioactive wastes are generated in many beneficial activities such as nuclear power production and a range of radioisotope application in medicine, industry, agriculture, and research.[17] Radioactive wastes are classified into low-, intermediate-, and high-level waste depending on their level of activity. The high-level waste (HLW) requires containment and isolation from humans and their living environment for a very long time.

Currently there are basically two options for managing this HLW. One option is to directly dispose spent fuel in a deep geologic repository to isolate it for the hundreds of thousands of years that it may remain hazardous. The other option is to reprocess the spent fuel to separate the uranium and plutonium for use as a new fuel. This second option, reprocessing, has been proven controversial. Proponents argue that a key, nearly unique, characteristic of nuclear energy is that used fuel may be reprocessed to recover fissile and fertile materials in order to provide fresh fuel for existing and future nuclear power plants. Indeed, a principal reason for reprocessing used fuel is to recover unused uranium and plutonium in the used fuel elements and thereby close the fuel cycle, and gaining some 25 per cent more energy from the original uranium in the process, and thus contributing to energy security. Furthermore, reprocessing reduces the volume of material to be disposed of as HLW to about one-fifth.[18]

On the other hand, opponents of reprocessing claim that it may be economically unfavourable, given that natural uranium is comparatively cheap. More importantly, the argument goes, separation, stockpiling, transport, and use of reprocessed nuclear fuel create risks of diversion to military purposes and risks of theft.[19] Given this uncertainty, most states have not publicly endorsed one option or the other. Instead they await further scientific advances and technological breakthroughs to determine the optimal choice.

In the United States, three civil reprocessing plants were built, but later closed for different reasons. The first plant at West Valley, New York, was operated successfully from 1966 to 1972. However, escalating regulation required plant modifications, which were deemed uneconomic, and the plant was shut down. The second plant built at Morris, Illinois, incorporating new technology which, although proven on a pilot scale, failed to work successfully in the production plant and was declared inoperable in 1974. The third plant at Barnwell, South Carolina, was aborted due to a 1977 change in government policy which ruled out all US civilian reprocessing as one facet of US non-proliferation policy. Since then the United States has not had been operating civil reprocessing plants.[20]

Similarly, the US government has not succeeded in choosing a repository to its radioactive nuclear waste. In 1987 the Congress selected Yucca Mountain in Nevada to bury the nation's nuclear waste. This choice was further confirmed in 2002. However, this choice was widely opposed by the people in Nevada, which does not have any nuclear plant, and their representative in the Congress. Furthermore, many people all over the nation criticized the choice on environmental and safety grounds. After spending $8 billion on the project, the Secretary of Energy Steven Chu announced in June 2009 that the administration decided to terminate the Yucca Mountain program and explore alternatives for nuclear waste disposal. In January 2010 the secretary established the Blue Ribbon Commission on America's Nuclear Future to recommend a new national strategy for managing spent nuclear fuel and HLW, including an examination of reprocessing and recycling options.[21]

Other nuclear power countries also have reprocessing and geological repositories. Britain, France, Russia, India, and Japan, among others, have established plants to reprocess HLW for decades. Russia has not yet chosen a final site for a geologic nuclear waste repository. Meanwhile, in recent years, Finland and Sweden have taken significant steps to establish geologic repositories. Posiva, a joint company formed by the two largest Finnish nuclear power utilities to take care of spent fuel disposal, is constructing an exploratory tunnel-to-disposal depth with a plan to apply for a repository construction license in 2012 so that final disposal can begin in 2020.[22] In a similar development, the Swedish Nuclear Fuel and Waste Management Company, owned by nuclear power plant operators, selected its repository site in June 2009 and expects operation to begin in 2023.[23] In November 2009 a giant step was taken by the launching of the Technology Platform for Implementing Geological Disposal of Nuclear Waste with the objective of coordinating the remaining research and development in the lead up to the construction and operation by 2025 of Europe's (and the world's) first deep geological repositories for high-level and long-lived radioactive waste.[24]

The non-existence of permanent geological repositories and the uncertainties regarding reprocessing have left many nuclear power states with few options – largely to store HLW in reactor pools close to the ground on a temporary basis. This interim stage allows effective cooling and radiation levels to decrease, thereby making handling of these radioactive wastes less hazardous. There is an international consensus that no existing or currently conceived future technology can eliminate the need for one or more geological repositories for long-lived radio nuclides. The good news is that permanent geological disposal does not need to occur immediately and technical options exist that can buy time for an incremental approach to repository development.[25]

To sum up, the world's decades-long experience with nuclear power suggests that a satisfactory solution to radioactive waste is yet to be found. Many countries have been debating different options for a long time. An optimal choice would require both technical and financial resources as well as political commitment and public awareness and participation. At present, there is no international mechanism for spent fuel disposal services. As Hans Blix, former head of the IAEA, said, the question of a final disposal plan 'is still open and more attention should be spent on deciding what to do'.[26] The most significant international treaty is the Joint Convention on the Safety of Spent Fuel Management and on the Safety of Radioactive Waste Management, formulated under the auspices of the IAEA. This first legal instrument to directly address these issues on a global scale was opened for signature in September 1997 and entered into force in June 2001. The convention applies to spent fuel and radioactive waste resulting from civilian nuclear reactors and applications and to spent fuel and radioactive waste from military or defence programs. At the end of 2010 more than 60 countries have signed the convention including Lebanon, Morocco, and the UAE.[27]

Challenge # 3: economic competitiveness

The recent global interest in nuclear power and the ambitious plans to build nuclear plants in many countries are likely to be restrained by the uncertainties regarding the economic competitiveness of this promising source of energy. Nuclear plants are quite capital-intensive. As discussed earlier, the ongoing efforts to improve their security have further added to their construction costs. At least three factors are likely to influence the cost-effectiveness and the broad calculus of the 'nuclear renaissance'.

Since the early 1950s nuclear power, natural gas, and coal (and to a less extent oil) have been considered the main fuels to generate electricity. The volatility of oil and gas prices and the rising concern about climate change due to pollution are a major drive for the switch to nuclear power. The development of new technology in the United States and elsewhere to explore and develop shale gas has drastically altered the dynamics of the global natural gas market. The International Energy Agency (IEA) projects that the glut in gas market and the drop in gas prices are likely to persist in the coming decades.[28] The availability of cheap natural gas supplies is likely to increase its attractiveness and competitiveness.

On the other hand, in order to curb emissions of carbon dioxide the international community in recent years has taken significant steps to progressively tax future carbon emissions at higher rates. Because nuclear plants do not produce carbon dioxide emissions, policies that place an explicit or shadow price on carbon dioxide emissions also affect their economic attractiveness compared to fossil-fuelled alternatives.[29] In other words, the economic competitiveness of nuclear power would be greatly enhanced if it becomes eligible for worldwide carbon trading schemes associated with the reduction of GHG emissions.

Worldwide supplies of uranium are abundant and secure. Uranium resources are generally considered to be sufficient for at least several decades. These resources are also highly diversified with uranium mining spread across the globe. According to the IAEA 19 countries are currently involved in uranium mining with eight countries (Australia, Canada, Kazakhstan, Namibia, Niger, the Russian Federation, Uzbekistan, and the United States) accounting for 93 per cent of world capacity.[30] This row uranium usually needs to be enriched. Most of enrichment facilities are concentrated in a few countries including Brazil, China, France, Germany, Great Britain, Japan, the Netherlands, Russia, and the United States.

Uranium prices are much less volatile than those of fossil fuels, though since the mid-2000s prices have increased substantially in anticipation of an increase in demand. This recent rise should not be seen as a major concern. A main advantage of uranium over fossil fuels is the small contribution of the former to the total cost of nuclear electricity.[31] Consequently, the case for building and operating nuclear plants is much less sensitive to variation in fuel prices than is the case for fossil-fuel-based generating plants.

This discussion of economic competitiveness of nuclear power and the global experience in the last several decades suggest two conclusions. First, the availability of uranium is apparently not a limiting factor at this stage of nuclear power development. Nonetheless, uranium resources, like fossil fuels, are not infinite and if the world's reliance on fission energy grows rapidly and sustainably, some means of extending uranium resources may become essential. Second, despite the many attractions nuclear power enjoys, it seems to go forward in most countries only where governments heavily subsidize its operations.[32] Reducing the cost of nuclear power (both capital and operating costs), while simultaneously improving safety, and waste management would enhance its competitiveness in a free market and reduce its dependence on government protection and subsidies and would pave the way for a large-scale nuclear energy growth around the world.

Challenge # 4: nuclear weapons proliferation

The renewed global interest in nuclear energy has heightened the concern over nuclear proliferation. The link between civilian nuclear power and nuclear weapons has been a serious unresolved worry throughout the nuclear age. The expansion of civilian nuclear power is likely to contribute to the dissemination of expertise, technology, and material that would be useful in launching a nuclear weapons program. Stated differently, the connection between civilian and military applications of nuclear power seems inevitable. The same fuel-cycle technologies – uranium enrichment and plutonium reprocessing – that are essential to power a nuclear reactor 'can, with adjustments, produce material to fuel the core of a nuclear explosive'.[33] This is dubbed the dual-use dilemma. Specifically, the challenge is that a potential spread of nuclear fuel cycle technologies, especially technologies for uranium enrichment and for reprocessing spent fuel to separate plutonium, poses a serious concern to the nuclear proliferation regime because enrichment and reprocessing capabilities give states the capability to produce fissile materials for weapons.[34] A commercial enrichment plant producing low-enriched uranium (LEU) could be reconfigured to produce high-enriched uranium (HEU). There is no technological barrier between the two processes.

The growing global interest in nuclear power has focused new attention on nuclear fuel production. The uranium enrichment market is currently dominated by a small number of large global companies such as European Gaseous Diffusion Uranium Enrichment Consortium (EURODIF), URENCO (multinational European consortia), Tekhsnabexport (Russian), and US Enrichment Corporation. This concentration has underscored the vulnerability of states with no national enrichment facilities and fuelled their suspicion that external supplies of enriched uranium may become associated with political pressure. The Iran-EURODIF affair is a case in point. In the 1970s the Shah of Iran loaned more than $ 1 billion to support the construction of the EURODIF diffusion enrichment plant in France. This gave Iran a 10 per cent share in the consortium and entitled the country to 10 per cent of the uranium enriched by EURODIF. With the 1979 revolution in Iran EURODIF refused to deliver the LEU to the Islamic Republic.[35]

This uncertainty regarding the supply of nuclear fuel has contributed to the controversy over the adequacy of the global regime to find the 'appropriate' balance between full utilization of nuclear power for peaceful applications and preventing nuclear weapons proliferation. In 1957 the international community established the IAEA with the mandate to pursue such balance, reassuring the international community that civil nuclear programs are not contributing to weapons acquisition. In the intervening decades, the role of IAEA has grown, evolving in response to the needs of member states. Early expansion in civilian nuclear power was accompanied by the development of nuclear applications in health, agriculture, hydrology, and industry.[36] After the uncovering of Iraq's clandestine nuclear weapons program in the 1990s, the IAEA's mandate was further strengthened by introducing the 'Additional Protocol', which gave the agency more inspecting power. The agency's growing importance was recognized by the awarding of the 2005 Nobel Peace Prize to the IAEA and its Director General.

The Non-Proliferation Treaty (NPT) is the key foundation of the global non-proliferation order. The treaty is based on a balance among three pillars: non-proliferation, disarmament, and peaceful uses of nuclear energy. These three pillars are integrally linked and the achievements in each area are likely to require progress in the others. The non-nuclear states insist on the priority of the second and third pillars, while the nuclear states emphasize the importance of non-proliferation.

The nuclear weapon states emphasize the objectives of preventing the further spread of sensitive nuclear technologies and material. They argue that the more states that acquire their own national enrichment or reprocessing capability, the more worrisome the nuclear future will be. On the other side, the non-nuclear weapons states insist that the NPT incorporated the principle that states have an 'inalienable right' to develop nuclear energy for peaceful purposes. Article IV states: 'Nothing in this Treaty shall be interpreted as affecting the inalienable right of all the Parties to the Treaty to develop research, production and use of nuclear energy for peaceful purposes'.[37] They further argue that it is a diversion to contend that the solution to non-proliferation rests in limiting technological rights for states that are in compliance with their treaty obligations.

This ongoing controversy raises two largely unanswered questions. First, will the non-proliferation regime be adequate in a world where there is more nuclear knowledge and technology spread across more states?[38] And second, does every country need its own nuclear fuel cycle? Or would it be more economical, with minimal risks of proliferation and an effective verification system, to rely on multilateral fuel banks?[39] It is apparent that the current international efforts to keep enrichment and reprocessing technologies concentrated in a few countries present a significant political challenge. Meanwhile, it is hard to deny the risks associated with widespread access to the know-how and material that could be diverted from intended peaceful use to military applications.

In order to solve this non-proliferation conundrum, several proposals have been considered since the dawn of the nuclear age. Their common goal is to prevent an increase in the number of states that would be capable of producing weapon-usable nuclear material. Most of them have sought to reach this goal by proposing that international entities take charge of uranium enrichment or platinum reprocessing instead of national authorities. In other words, the internationalization of the nuclear fuel cycle is their underlying theme. Most of these proposals are generally considered technically feasible, but lack political consensus and commitment, particularly among nuclear power aspiring states.

One of the earliest proposals – the Report on the International Control of Atomic Energy (also known as Acheson-Lilienthal Report) – was introduced by the US government in March 1946. Its primary message was that control of atomic energy through inspection and policing operations was unlikely to succeed. Instead, the report proposed that all fissile material be owned by an international agency to be called Atomic Development Authority, which would release small amounts to individual nations for the development of peaceful uses of atomic energy.[40] The lack of trust, indeed hostility, between the then world's two superpowers and the general dynamics of the Cold War doomed the proposal. These dynamics prevailed in the ensuing decades until the dissolution of the Soviet Union in the early 1990s.

Since the early 2000s the renewed global interest in nuclear power and the fact that several aspiring states are considering establishing their own national nuclear fuel cycle facilities have opened the door for more proposals to control nuclear technology and material. Austria, Germany, Japan, Russia, the United Kingdom, and the United States, among others, have introduced similar proposals. One of the few proposals that originated from non-nuclear supplier states was an initiative by the Gulf Cooperation Council (GCC) – Bahrain, Kuwait, Oman, Qatar, Saudi Arabia, and the UAE. In October 2007, the GCC put forward an initiative that invited all interested states of the Middle East to participate in the establishment of a uranium enrichment international consortium, to be based in a neutral country outside the region. The proposal sought to stop Iran from enriching uranium. Tehran did not reject the proposal, but insisted on maintaining its enriching facilities.

In December 2010, after years of negotiations, the IAEA approved a nuclear fuel bank that nations can draw on for their atomic programs in emergencies. This fuel bank will ensure the sale of uranium for power plants to countries that are in good standing with the IAEA. It is meant to be a backup in case countries face a cut-off from commercial suppliers. The fact that it is run by the IAEA means a further guarantee of its impartiality.

The majority of non-supplier states have been suspicious of most of the proposals to curb the dissemination of nuclear technology. They tend to see them as technology-denial measures and attempts by the established nuclear "haves" to uphold the technological edge.[41] Furthermore, they are reluctant to accept additional restrictions on nuclear activities beyond those already in place under the NPT. Finally, they question the reliability of nuclear fuel supplies based on these proposals.

To sum up, multilateral proposals to the nuclear fuel cycle are 'work in progress' that requires close cooperation between the states that already have the technology and capability and those who aspire to have them. A central problem hampering progress is the lack of trust between the two sides. The IAEA, which is likely to play a leading role in closing the gap between the two sides, seeks to assure all countries in compliance with their obligations to the non-proliferation regime that they will be able to obtain reliable and reasonably priced supply of fuel cycle services.

This brief discussion of safety, waste management, economic competitiveness, and weapons proliferation are among the most serious challenges facing the global 'nuclear renaissance'. These challenges, however, have not restrained the growing interest in nuclear power by both well-established nuclear states and newcomers. Establishing a nuclear power program in the Middle East (and elsewhere) requires developing technical and human infrastructure, regulatory framework, enacting appropriate laws, promoting safety culture, adhering to relevant international agreements and treaties, allocating the necessary funding, securing nuclear fuel, and articulating a policy on nuclear waste.[42]

The underlying fact is that nuclear power, like all other sources of energy, has its own advantages and disadvantages. Nuclear power, while significantly more attractive today than a few decades ago, still has unresolved challenges related to safety, spent fuel, nuclear proliferation, and economics. It should not be seen as a panacea. Rather, the growing global concerns and efforts to address energy security and climate change should have realistic expectations and see this promising nuclear power as an essential component of diversified energy mix. Another important part of this mix is renewable energy.

Renewable energy

Like nuclear power, renewable energy sources have been the topic of continued interest in both developed and developing countries. The reasons are similar to those behind the 'nuclear renaissance' – energy security and climate change. Headlines announcing gas supply cuts to the Ukraine, oil tanker hijackings along the coast of Somalia, pipeline bombings in Nigeria, threats to close Strait of Hormuz, and hurricanes destroying oil rigs in the Gulf of Mexico showed that threats to energy security arise in many forms and in unexpected places. Meanwhile, the growing interest in slowing climate change and reducing pollution has intensified global efforts to reduce dependence on fossil fuels and search for alternative sources.

Renewable energy is any form of energy that is replenished by natural processes at a rate that equals or exceeds its rate of use. Renewable energy is obtained from the continuing or repetitive flow of energy occurring in the natural environment and includes resources such as biomass, solar energy, geothermal heat, hydropower, tide and waves, ocean thermal energy, and wind energy.[43] Some renewable energy resources such as hydropower are technically mature and have been deployed at a significant scale. Others, such as wind, solar, and geothermal, are in a nascent phase of technical maturity and commercial production and deployment. Unlike fossil fuels, almost all countries have access to some forms of renewable energy. For example, solar and ocean energy are widely distributed. Still, the contribution of renewable energy to the overall energy mix varies substantially from one country to another.

Against this background, renewable energy has been expanding rapidly in recent years. With the exception of hydropower, most renewable sources start from a small base. The greatest scope for increasing the use of renewables lies in the power sector. The IEA projects that renewables-based generation will triple between 2010 and 2035 and the share of renewables in global electricity generation will increase from 19 per cent in 2010 to almost one-third during the time span.[44] Similarly renewable energy is projected to supply 7 per cent of world transport fuels.[45] This current and projected surge of renewable energy illustrates the many benefits it can provide. These include the reduction of carbon dioxide and other GHGs, enhancing energy security by diversifying the energy mix and making energy accessible, particularly in rural areas. Little wonder, renewable energy has attracted considerable attention and investment in recent years. Thus, by 2010 renewable energy accounted for $112 billion of global investment, up from just $18 billion in 2004.[46] Equally impressive, 2011 was the first time global investment in new renewable power plants surpassed fossil-fuel power plant investment.[47]

Despite this impressive progress, it is important to point out that the development and deployment of renewable energy face significant challenges and uncertainties due to the European economic crisis, which led several governments to cut their financial support. Equally important, the shale gas boom in the United States has pushed gas prices to very low levels. Sustainable low gas prices would make renewable resources less competitive. In the Middle East, the renewable energy industry has to overcome several hurdles including lack of adequate national institutions, conflicting legal and regulatory frameworks, political instability, and uncertain investment environment.[48]

A number of characteristics can be identified in the majority of renewable sources. First, renewable sources are largely site-specific in many, if not all, their applications. Unlike oil, natural gas, and coal, which are transported across the globe with different degrees of ease, renewable sources are more likely to be consumed locally. Some analysts see this locality as an advantage particularly as the technology advances and prices drop. For example, wind and solar energy can provide electricity to remote and isolated communities. Second, in many cases renewable sources supply energy intermittently. The sun and winds are not uniform in their availability for power generation on a daily basis and vary from one season to another, and hydropower depends on rainfall. There are ways to address and at least partly overcome this intermittency issue such as by the creation of a back-up system to fill the gap when renewables fail to deliver. Still, the intermittency issue makes renewables more expensive and less reliable.[49]

Third, despite renewable energy's environmental advantages, it is largely more expensive than fossil fuels and still provide a small share of the global energy mix. Though renewable energy is becoming more economically competitive, substantial investments and technological advances are needed. To a great extent, renewable sources cannot compete with oil, natural gas, and coal in a free market in many countries. They need government protection and subsidies. Fourth, renewable energy sources differ in many of their important characteristics, costs, and availability. It is too simplistic to lump all of them together in a single category. A brief survey of the major renewable sources highlights the similarities and differences between them.

Hydropower

Hydropower is a major and one of the oldest forms of renewable energy where power is derived from the energy of water moving from higher to lower elevations. The mechanical power of falling water is an old tool used for a variety of services for thousands of years. Historically, most renewable electricity has come from hydropower. In the twenty-first century, hydropower is considered a mature, predictable, and price-competitive source of energy. It generates approximately 16 per cent of the world's total electricity and 86 per cent of all electricity from renewable sources.[50]

Hydropower's dominant contribution to electricity generation is due to several characteristics. These include large capacity range, flexibility, storage capability (when coupled with a reservoir), and ability to operate in a stand-alone mode or in grids of all sizes. These characteristics make hydropower particularly valuable to address the fluctuations in electricity demand and to match supply from less flexible and less reliable renewable sources.[51] Finally, hydropower is one of the least pollutant sources of energy.

This promising current and projected hydropower's potential should be balanced by three caveats. First, construction costs for new hydropower projects vary depending on the unique nature of each one. Parameters affecting investment costs and the return on investment include the project scale, project location, the presence and size of reservoir, and the other possible benefits such as flood control, irrigation, and freshwater supply. Thus, the costs of developing and operating hydropower projects vary widely from one project to another.

Despite this variation, hydropower is generally economically competitive. Second, potential environmental impacts need to be anticipated and appropriate steps taken in advance to avoid, mitigate, or compensate for them. The construction of hydropower projects can affect air, land, water, and biodiversity of surrounding communities. Some of the prominent impacts include changes in river flow, water quality and quantity, barriers to fish migration, and loss of biological diversity. Third, similarly, social impacts need to be taken into consideration. These include protecting the lives and property of citizens from floods and/or droughts and securing citizens' rights regarding expropriation of land to be inundated. Local citizens, the stakeholders most impacted by hydropower projects, need to be fully informed and consulted.

To sum up, while hydropower is a well proven and mature technology, it is still advancing and expanding its scope of application. Indeed, the IEA projects that hydropower will continue to be one of fastest growing sources of energy.[52] It addresses global and national concerns over climate change and energy security. In order to initiate and expand hydropower projects, environmental and social impacts need to be explored and appropriately addressed and mitigated. The International Hydropower Association (IHA) was founded in 1995 to address the role of hydropower in meeting the world's growing water and energy needs as a clean, renewable, and sustainable technology.[53]

Geothermal

Geothermal resources consist of thermal energy from the earth's interior stored in both rock and trapped stream or liquid water. The tapped heat from an active reservoir is continuously restored by natural heat production, conduction, and convection from surrounding hotter regions, and the extracted geothermal fluids are replenished by natural recharge and by reinjection of the cooled fluids.[54]

Geothermal energy enjoys several advantages in comparison to other sources. Unlike fossil fuels, geothermal energy is infinite and produces no pollutants from emissions. Compared with other renewable sources, it also reflects different characteristics. People who object to wind farms, bio-fuel, or photovoltaic systems have no reason to make similar claims over geothermal energy since it is 'more or less invisible, inaudible and odourless'.[55] In short, geothermal energy has the potential to provide long-term, secure base-load energy with little environmental footprint. Given these advantages, geothermal energy has been utilized for more than a century. Thanks to technological progress the global production of geothermal electricity has grown rapidly in recent years. It is particularly used in providing heating and cooling for buildings, greenhouses, bathing, swimming pools, and water desalination.

Despite geothermal energy's potential and advantages, its use is limited to countries and regions with considerable volcanic activity (which allows access to very high-temperature fields). In the Middle East, Iran is one of a few countries with relatively developed geothermal energy. More specialized research and advanced technology are needed to reduce costs and improve security and prolong the life of production facilities. Furthermore, geothermal projects are characterized by long lead times and a high degree of exploration risk. Therefore it is not surprising the IEA projects that the geothermal sources would provide a small share of renewable generation in the foreseeable future.[56]

Bio-fuels

A bio-fuel is a solid, liquid, or gaseous fuel derived from biomass. The solids that can be used as fuels through combustion are wood, straw, dried plants, cattle dung and other animal droppings, husks, and bagasse. The liquids are ethanol, which is mainly produced from sugarcane and corn, and biologically produced oils that can be used in diesel engines. The gaseous source of fuel is methane produced by decaying garbage or manure in places from where it can be collected.[57]

The initial hyper-excitement about bio-fuels has waned. While some analysts and policymakers continue to believe that bio-fuels can contribute to GHG emissions and enhance energy security, others see several drawbacks that need to be addressed. These include the impact on food prices due to competition with food crops, high production costs, limited GHG reduction, negative impact on deforestation and biodiversity, and pressure on scarce water resources.[58]

The cumulative impacts of these concerns have increased the interest in developing bio-fuels produced from non-food biomass. Feedstocks from lingo-cellulosic materials include cereal straw, bagasse, forest residues, and purpose-grown energy crops such as vegetative grasses and short rotation forests.[59] Despite these concerns and limitations, bio-fuels have steadily grown since the early 2000s and are projected to maintain high level of growth in the foreseeable future. This growth, however, depends on the ability of modern technology to bring costs down. Like other renewable sources, the costs of bio-fuels vary from one country to another based on feedstock types and the scale and timing of production. Currently bio-fuel production is largely concentrated in

Brazil, the United States, and the European Union with limited trade on the global markets.

Wind

Wind power is one of the oldest sources of energy in human history. It has been used for thousands of years in a wide range of applications. Technological progress has made it possible to generate electricity from wind since the 1970s. Since then wind power has become one of the most mature sources of renewable energy. It is a major source of power in over 70 countries across the world. Between 2003 and 2010 wind generation more than tripled. Most of this growth took place in the United States, Europe, and Asia. In late 2000s the United States overtook Germany and became the world's leader in wind power generation. On the other hand, nearly 20 per cent of Denmark's electricity generation comes from wind. The success of wind power is not confined to Western countries. In 2009 China became the world's second largest installer of wind power and the largest manufacturer. Similarly, India has a strong wind industry, and rapid developments are also taking place in Africa and Latin America.[60] Wind power is also a mature and growing industry in several Middle Eastern countries. In late 2000s the annual growth rate in Iran was 7.1 per cent, Egypt 17.8 per cent, Turkey 74.9 per cent, Morocco 88.8 per cent, and Tunisia 170 per cent according to the Global Wind Energy Council.[61] Despite this growing interest in wind power, it is important to point out that wind potentials are not evenly distributed between regions and countries. Furthermore, wind farms are usually far away from major population centres.

Wind power comes from onshore and offshore farms. The onshore wind is well-established industry and is considered one of the most cost-competitive renewable energy sources. In the last decade it experienced a 27 per cent average annual growth rate and is projected to continue this impressive expansion.[62] On the other hand, offshore wind is still considered an emerging industry that requires further research and development to enhance the technology and reduce costs.

Wind turbines both onshore and offshore are complex systems that require considerable studies and planning to choose the most appropriate location and equipments in order to optimize costs and performance. More studies are needed to determine the impact of wind plants on local climate, wildlife, and the broader ecosystem. Similarly, more efforts are needed to establish regulatory systems and new institutions, particularly in countries that have not yet experienced substantial wind energy deployment (like most of Middle Eastern states). Sharing experience and technology with these countries which have recently launched plans to utilize their wind power potentials would help in the installation and operation of new plants.

Despite this growth, the world still generated less than 1 per cent of its total electricity from wind power in 2008. Wind power is projected to account for more than 3 per cent of total world electricity by 2020.[63]

Solar

Solar power utilizes the energy from sunlight either directly or indirectly. It has been used for heating and cooling, generating electricity, water desalination, and many other residential, commercial, and industrial applications. Solar power offers some substantial advantages over other energy sources. Solar-generating facilities are most productive in the middle of the day, when demand for electricity is at its peak. Unlike fossil-fuel-fired generating capacity they produce no toxic emissions and unlike nuclear plants they leave no atomic waste. Roof-top solar panels can be installed in homes and businesses, reducing the need for centralized power plants and transmission lines. And, most importantly, the sun's rays are free and available in infinite quantity.[64] Additionally, solar power technologies use a small amount of land in comparison with other sources of energy. Solar power does not require mining, transport, or disposal. Indeed, the land used to generate solar power is often infertile and unused. Therefore, solar energy is the most abundant of all energy resources. Although it is not equally available in all countries, solar power has the potential to make a significant contribution to the energy mix of almost all countries and regions.

Generation of electricity from solar can be achieved mainly in two ways: photovoltaic (PV) systems and solar thermal. PV technology involves the generation of electricity from light by using a semiconductor material that can release electrons, which in turn form the basis of electricity. All PV cells have at least two layers of such semiconductors (positively charged and negatively charged ones). Solar thermal technologies or concentrated solar power (CSP) concentrate energy by heating up water in a dark vessel. The heat is then transferred to operate a conventional power cycle.[65]

In recent years solar PV has been the fastest-growing renewable energy technology worldwide. This growth, however, has been concentrated in only a few markets (Germany, Italy, Japan, and the United States). Progress in CSP has lagged behind, with Spain and the United States taking the lead. In the Middle East several countries including Iran, Saudi Arabia, the UAE, Israel, Egypt, and Morocco (among others) have launched ambitious programs to utilize their solar power potential. Indeed the entire Middle East region is amply blessed with enormous solar power potential. Generally the Middle East receives an average daily sunshine of about nine hours and has little rainfall.[66] The region's climatic conditions are highly conducive for developing solar power on a large scale. A recent study by the German Aerospace Center concluded that the Middle East region's solar potential 'far exceeds global electricity demand'.[67] These resources, however, have been largely untapped.

Solar power faces a number of challenges and hurdles particularly with regard to land requirements for centralized PV and CSP plants and for the latter, cooling water requirements. Choosing low population density areas and using dry cooling approaches have helped overcoming these hurdles. The future expansion of solar power will depend on the continuation of supportive public policy, technological innovation, and cost reduction. In the last few decades the industry has witnessed substantial cost reduction.

2
Morocco

The kingdom of Morocco is different from its North African neighbours in many ways. Algeria, Libya, and Egypt hold substantial oil and gas deposits and for a long time have been its major producers and exporters. Morocco, on the other hand, has very limited indigenous reserves. The kingdom is heavily dependent on foreign supplies to meet its growing energy demands. The country's high rates of population growth and urbanization, as well as economic expansion, mean rising energy consumption and wider gap between supply and demand. This heavy dependency on imported fossil fuels adds more burden on the country's trade and financial deficit.

This energy outlook, however, is not completely bleak. Indeed, there is room for optimism. Morocco enjoys significant potential to meet its energy needs from renewable sources, particularly hydropower, solar, and winds. The country is blessed with several rivers including major ones such as the Sebou, which runs into the Atlantic, and the Moulouya, which runs into the Mediterranean. Large part of Morocco is desert with sunny days almost all year round. This desert is also sparsely inhabited. These two characteristics give Morocco great advantages to become a leading source of solar power. In addition, Morocco's long coast lines and mountains give it significant potential to utilize, largely untapped, wind power. Finally, Morocco enjoys a strategic location as the closest connection between Europe and Africa. This geographical proximity, among other reasons, has laid the foundations for energy partnership between Morocco and the Europe. Indeed, Morocco is the only African country with a power cable link to Europe.

The Moroccan government has been aware of the country's severe shortage of fossil fuel deposits and rich renewable energy potentials. In the last several years the government has been under pressure to cap the bill for energy, more than 90 percent of which is imported.[1] The Ministry of Energy, Mines, Water and Environment is in charge of the energy sector.[2] Working with other authorities, the ministry announced a National Energy Strategy in 2009. The strategy focused on security of supply, diversification of energy mix, accessibility of energy for all (e.g. rural and urban population), low cost, safety, efficiency, and clean environment. The strategy sought to increase the share of renewable energy to 10 percent of primary energy supply and to 18 percent of power generation. Finally, the strategy called for intensifying efforts in oil and gas exploitation.[3] At the same time, it sought to raise the share of natural gas in power generation from 5 per cent in 2009 to 30 per cent in 2030. This rise will be at the expense of coal (to fall from 34% to 19%) and oil (to fall from 24% to 8%) during the same time span.[4] In short, the strategy seeks to achieve a two-fold objective – increase supplies and diversify the energy mix, with particular focus on renewable sources, and restrain consumption by improving efficiency. The Moroccan Agency for Solar Energy (MASEN) stated the main objective of the country's energy policy:

- To strengthen the security of supply through diversification of the energy mix
- To provide access to modern energy at competitive prices for the entire population
- To optimize costs through gradual market liberalization and in-depth reforms of the energy sector
- To control consumption and enhance energy efficiency and saving[5]
- To promote renewable energy and clean energy technology
- To protect the environment and limit the greenhouse gas emissions
- To support regional integration through the opening up to Euro-Mediterranean energy markets and the harmonization of the energy regulations and legislations.[6]

The electricity sector illustrates the challenges (and opportunities) facing the broader national energy strategy. For a long time the rural population did not enjoy the convenience of electricity available in major urban centres. In 1995 the rate of rural electrification was only 18 per cent. In 1996 the National Electricity Office (ONE) launched a national electrification program known as Programme pour L'Electrification Rurale Global. By the end of 2000s the rate rose to close to 100 per cent.[7] This success underscores the approximately 8 per cent annual increase in electricity demand nationwide.

Fossil fuels, by far, dominate the electricity sector, with a small share provided by hydropower. The country's two largest electricity power stations at Mohammedia and Jorf Lasfar are both coal-fired. Morocco produces a small and declining amount of coal from a mine at Jerada, but most of the coal is imported from South Africa. In recent years the government has sought to reduce this heavy dependency on fossil fuels and replace them with renewable sources. In late 2000s Rabat announced plans to expand electricity generation from wind and bring the share of total generation power from winds to 12 per cent by 2015.[8]

Until recently, the government dominated almost all the operations in the electricity sector including generation, distribution, and pricing. In the last few years private sector and foreign companies have been allowed to play a bigger role. Thus, Jorf Lasfar became Morocco's first privately operated power station in 1997, when it was taken over by a US–Swiss consortium. In the following years ONE signed contracts with Temsol, Total Energie, and Apex-BP, among others, to supply electricity to rural areas. Equally important, subsidies on electricity prices have been phased out in recent years.

To sum up, population growth and economic prosperity are pushing energy consumption upward. Meanwhile, Rabat has very limited hydrocarbon deposits and has increasingly sought to utilize its renewable source potential as well as improve energy efficiency. In pursuing this broad strategy, the government has sought to partner with the private sector and foreign companies and reduce its control over the energy sector. The success of this strategy is likely to reduce pollution and improve the country's trade and budget deficits. The following sections provide a detailed analysis of this broad energy strategy.

Fossil fuels

Exploration for oil in Morocco started in late 1920s and the volume of production increased slowly in the following decades, but never matched those of North African neighbours. Technological advances and the discovery of oil deposits in Algeria and Mauritania have raised the expectations that similar reserves might be found on the Moroccan side. Currently there are a few places with proven reserves such as the Rharb Basin, Essaquira, Prerif, and Tarfaya.[9] The active oil and natural gas fields are found in the Essaquira Basin on the coast (both oil and natural gas), the Gharb Basin in the north (natural gas), and Meskala (natural gas).[10] Morocco imports a large proportion of its natural gas needs from neighbouring Algeria via the Meghreb-Europe Gas pipeline (MEG), also called

Pedro Duran Farell. Originally, the pipeline was proposed in the early 1960s, but the dispute over the Western Sahara prevented the parties from reaching an agreement. Three decades later, the pipeline was built and came online in 1996. It connects Algeria's gas field Hassi R'Mel with Cordoba in Spain via Morocco. It ties into the Spanish and Portuguese natural gas transmission networks. An international consortium, led by Spain's Enagas, Morocco's SNPP, Portugal's Transgas, and Algeria's Sonatrach, operates the pipeline.[11]

The two government institutions in charge of developing and marketing the country's oil and gas resources are the national oil company, Office Nationale de Recherches et d'Exploitations Petrolieres (ONAREP), and the National Office of Hydrocarbon and Mines, Office National Des Hydrocarbures Et Des Mines (ONHYM). They work with national and international companies to meet the nation's demand for fossil fuels. In the mid-2000s China Offshore Oil Corporation, Norway's Norsk Hydro, and Denmark's Maersk signed agreements to explore for oil and gas. In addition, the Moroccan government gave concessions to several international companies to explore for oil in the Western Sahara. The list includes Total, Wessex Exploration, Svitzer, Wales Robertson Research International, and TGS Nopec. ONAREP and ONHYM activities include exploitation, research and development, and downstream. Despite limited indigenous crude deposits, Morocco has well-developed refining industry. The two main refineries are the Societe Anonyme Marocaine de L'Industriele Raffinage at the port of Mohammedia and the Societe Cherifiennedes Petroles at Sidi Kacem.

In addition to 'traditional' oil, oil shale has the potential to contribute to Morocco's energy security. The country's oil shale reserves are largely untapped and unutilized. Exploration for the recovery of oil shale started at Tangier with the creation of the Societe des Schistes Bitumineux de Tanger. The company built a pilot plant with a production capacity of 80 tons per day of oil shale between 1939 and 1945.[12] The technological advances in the oil shale exploration and the surge in oil prices in the 1970s added more interest in the already known reserves in Morocco. Particularly the deposits found near Timahdit (southeast of Rabat) and Tarfaya (southwest part of Morocco) have received special attention. These deposits were originally discovered in the 1960s. Several energy companies from North America and Europe conducted exploratory drilling and experimental mining and processing of Moroccan oil shale.[13] Following these tests the ONHYM developed and tested a shale oil extraction process. Still no shale oil was produced.

Since the mid-2000s several international oil companies have signed agreements with the Moroccan authorities to further explore and develop oil shale deposits in Timahdit and Tarfaya. The list includes Total S.A., San Leon Energy, and Eesti Energia.

To sum up, although Moroccan oil shale reservoirs have been studied and tested since the 1930s, there is no commercial production yet. Such projects involve an enormous quantity of rocks and necessitate major investments. The existence of both technological and financial risks has created a road block to the development of oil shale deposits. Finally, it is important to point out that the processing of oil shale generates by-products (e.g., sulphur, gas, and ammonia), which have a real industrial and commercial importance. The retorting and combustion ash could be used for clinker production in the cement manufacturing industry. The utilization of these valuable by-products is what will make oil shale development worthwhile and economically feasible in the future.[14]

Nuclear power

Morocco's reasons to initiate a nuclear power program are not different from other Middle Eastern countries. As stated above the country's electricity consumption is surging and it lacks indigenous energy resources. The country's long shores on the Mediterranean Sea and the Atlantic Ocean mean substantial amount of sea water can be used after desalinization. Desalinization is an energy-intensive process. Thus, the gap between energy supply and demand is large and growing. Reliance on imported supplies adds more burden on the country's economy. Finally, rising industrialization and growing urbanization contribute to mounting pollution.

Against this background one can understand Morocco's interest in nuclear power. Lacking any technical infrastructure, Rabat has sought foreign partners and signed agreements with a number of foreign companies and governments. In the late 1990s it carried out a feasibility study for a Chinese-built 10 MW demonstration plant at Tan-Tan on the Atlantic coast with technical assistance of the International Atomic Energy Agency (IAEA) and financial backing from the European Union.[15] In October 2007 a partnership with France to build a nuclear power plant near Marrakesh was established and a nuclear energy cooperation agreement was signed. Addressing the Moroccan authority, French President Nicolas Sarkozy stated, 'France will be your partner, France is making this political choice and sill support you on this path'.[16]

Paris and Rabat signed another cooperation agreement in 2010. France's Prime Minister Francois Fillon stated that the agreement was not a commercial deal to build a nuclear reactor, but rather a 'framework accord that will help Morocco prepare its entry into the field of nuclear energy'.[17] The French company Areva signed accord with the Moroccan authority to investigate recovery of uranium from phosphoric acid. The Russian company Atomstroyexport signed an agreement to build a nuclear power plant at Sidi Boulbra and the US company General Atomics signed a deal to build a research reactor east of Rabat.

The 2009 Copenhagen Accord invited developing countries to submit proposed Nationally Appropriate Mitigation Actions (NAMAs) demonstrating how they planned to reduce their greenhouse gas emission through specific proposals. Under these terms Morocco announced plans to build two 1000 MW nuclear reactors. In addition, the authorities approved plans to set up a nuclear safety agency and enacted a law on nuclear security.[18] Finally, Morocco joined several international initiatives and organizations to ensure the safety of nuclear reactors and materials and to combat nuclear terrorism.

Renewable energy

Morocco's pursuit of nuclear power will take some time and, like other sources of energy, nuclear power is not a panacea. Rather, issues such as fuel cycle and nuclear waste (among others) need to be addressed. At the same time Rabat lacks indigenous hydrocarbon reserves and the high consumption of imported oil and coal has significant environmental impact in terms of greenhouse gas emissions. In rural areas, where biomass is used for heating and cooking, the pressure on forests is high. In short, Morocco has many good reasons to utilize its substantial renewable energy potentials.

At least since the early 1980s the Moroccan authorities have shown great interest in developing solar and wind energy as well as the relatively developed hydropower. In February 2008 the government launched the National Renewable and Efficiency Plan with ambitious goals to increase the share of renewable energy in the national electricity balance by 20 per cent and their contribution to the energy mix by 10 per cent. The plan calls for achieving these ambitious goals by 2020.[19] The plan is projected to create more than 40,000 jobs and stimulate over $4.5 billion in investment.[20] In January 2010 the Moroccan Parliament adopted a new legislation for the promotion of renewable energy deployment. The law included provisions for authorizing electricity generation from renewable energy sources by private companies and the establishment of a new agency for renewable energy and energy efficiency (ADEREE) to replace the existing Renewable Energy Development Center (CDER).

The CDER was established in 1982 with a responsibility to educate the public and train specialists in the area of renewable energy. In partnership with the United Nations Development Program (UNDP), it provided financial and technical assistance to approximately 200 Maisons de L'Energie. These maisons (houses) are small rural enterprises that help local residents to produce their own energy supplies.[21]

The ADEREE is better funded than its predecessor and has broader responsibilities. It is in charge of implementing the nation's National Plan for Renewable Energy and Energy Efficiency. Specifically, its responsibilities include the following:

- Proposing national and regional plans and incentive measures for renewable energy and energy efficiency
- Development of programs for renewable energy and energy efficiency within the framework of the national energy strategy
- Identification, evaluation, and realization of the cartography of renewable energy resources and ideas for energy efficiency as well as proposing development zones for solar and wind energy
- Developing standards in the fields of renewable energy and energy efficiency as well as supporting research and development and training
- Follow-up and coordination at national level of energy audits
- Promoting technical and financial cooperation at national and international levels.[22]

Hydropower: Taking advantage of its geography, the Moroccan authority has generated electricity from dams for more than a century. Some of the oldest dams include Sidi Said Machou (1929), Fes Aval (1934), Zidania (1935), and Lalla Takerfoust (1938). More recent ones include Idriss the First (1978), Oued El Makhazine (1979), Al Massira (1980), Allal el Fassi (1994), and Al Wahda (1997).[23] One of the gigantic project combines hydropower and wind power in one single location. In 2012 the project is in the planning stage and upon completion it will double the existing hydropower capacity of Morocco.[24] The total cost of the project is about $2.16 billion. The Clean Technology Fund and the African Development Bank are among the top investors.[25]

Wind: Morocco is one of the leading wind power nations in North Africa. The wind is generally highest at the northern and south-western coast and on the eastern part of the Atlas Mountains. The areas with substantial excellent wind energy potential include the Essaouira, Tanger, Tetouan, Dakhla, Tarfaya, and Taza. In recent years substantial efforts have been made and massive financial resources have been allocated to develop wind energy.

Since the early 2000s Morocco has built several wind farms.[26] El Koudia El Baida, installed in 2000, was the first wind farm in the country with a capacity of 50.4 MW. Compagnie Eolienne du Detroit, a subsidiary of the French company Tholia, was in charge of constructing the farm. It generates about 1 per cent of the national annual electricity consumption. Lafarge Wind Farm was built between 2005 and 2009 by the cement company Lafarge for the electricity supply of its factory near Tetouan. Any excess power not consumed by the Lafarge factory is fed into the national grid. Amogdoul Wind Farm came online in April 2007. It is located in Cap Sim south of Essaouira. The biggest and most recent one is Tangier Wind Farm. It was commissioned in April 2009 and additional capacity was added the following year. In a sign of strong official support, King Mohamed VI chaired the inauguration ceremony.[27]

In 2010 the government launched the Moroccan Integrated Wind Energy Project with an estimated investment of $3.7 billion. The project aims to raise the installed wind capacity, thereby increasing the share of wind power in the national energy mix. The realization of these goals would contribute to reduction in imported fossil fuels and in greenhouse emissions. As part of the Integrated Wind Energy Project, the authorities have proposed the construction of several wind farms with a total capacity of 850 MW. They are Tanger II (150 MW), Midelt (100 MW), Jbel Lahdid (200 MW), Tiskrad (300 MW), and Boujdour (100 MW).[28]

To sum up, Morocco is blessed with huge wind energy potentials. The authorities have already invested large resources in developing these potentials and made an impressive progress. Still, the full utilization of these potentials requires billions of dollars and advanced technology. In the last several years the Moroccan government has sought to provide financial incentive and create an attractive investment environment by relaxing its control over the energy sector and establishing a partnership with the private sector and foreign companies.

Solar: Morocco's abundant sunshine, low humidity, and large tracts of unused land near road networks and power lines make it an ideal place for solar energy in general and for concentrated solar power plants in particular. In recent years the Moroccan authorities have focused considerable attention on promoting solar power.

In the mid-2000s the National Electricity Office (ONE) developed a program for the electrification of areas by utilizing solar power. The program was a joint venture with foreign companies including Total and Tenesol. The challenge was that in regions where homes are scattered, it is impossible for the electricity grid to reach each individual home in a cost-effective way. In order to generate its own electricity, each home was fitted with a solar system in which the solar panel turns the sun's rays into electricity, which is stored in a solar battery that provides electricity at night. The battery stores enough power to last up to five days. The electronic controller automatically manages the charging and discharging of the battery.[29] Specifically, the project's objectives were as follows:

- To provide affordable and sustainable rural electrification
- To improve rural local conditions to decrease rural–urban migration
- To develop rural economies
- To increase use of renewable energy.[30]

In 2009 Morocco launched its most ambitious solar program, the Solar Plan Initiative. It seeks to enhance the nation's energy security by diversifying the energy mix, promote productivity and competitiveness in the energy sector, protect the environment, and consolidate cooperation with the European Union (EU). This $9 billion initiative includes the installation of 2000 MW in five sites by 2020. These sites are: the 500 MW Ouarzazate (to be commissioned in 2015), the 400 MW plant next to Ain Beni Mathar (to be commissioned in 2016), a 500 MW plant in Foum Al Ouad (to be commissioned in 2017), a 500 MW plant in Boujdour (to be commissioned in 2018), and a 100 MW plant to be commissioned in Sebkhat Tah (to be commissioned in 2019).

The location of the Ouarzazate plant in eastern Morocco was chosen based on several considerations. It has one of the highest sunshine levels in the world, next to paved roads, close to Mansour Eddahbi dam (access to water is important for running concentrated solar power plants) and in a low seismic risk region. Furthermore, the site is far from any protected natural or touristic areas and away from any population centres. In other words, the project will cause very few land-use conflicts because it is currently used as grazing land with very little pasture. No displacement of communities or economic activities is projected.[31]

These plants will use cutting-edge design, combining a large array of parabolic mirror collectors concentrating sun energy and boosting the steam output needed to produce electricity. They will produce low-carbon solar electricity, create jobs to local communities, establish training centre and technical expertise, and pave the way for the deployment of a pipeline of solar energy projects in Morocco and across North Africa.[32]

According to Amina Benkhadra, Minister of Energy, Mining, Water, and Environment, the production capacity represents 38 per cent of the installed power by late 2008 and 14 per cent of the electric power by 2020. The project will significantly contribute to the government's efforts to reduce pollution. Upon completion, it will save approximately one million ton oil equivalent (TOE) and avoid the emission of 3.7 million tons of greenhouse gas annually.[33] The funding for this huge project is based on public-private partnership (PPP). The World Bank' Clean Technology Fund, International Bank for Reconstruction and Development, African Development Bank, Instituto de Credito Official (Spain), and the Moroccan state-owned National Electricity Office are among the main investors.[34] The project will be implemented in four phases.

The Moroccan Agency for Solar Energy (MASEN) was founded in 2010 to oversee all aspects of the Solar Plan Initiative. Its mandate is to implement the overall project (design, choice of operators, implementation, and management) and to coordinate and supervise all other activities related to the project.[35] MASEN is also responsible for developing national expertise by sponsoring educational and training centres.

Some of the main challenges facing the Solar Plan Initiative include the shortage of indigenous professionals and skilled labourers and funding uncertainties.[36] However, the country's power interconnections and integration within the EU energy space is likely to attract investment and provide opportunities to export electricity. Finally, Morocco's solar plants are being closely watched in Africa, where some hope that the poor energy infrastructure in many countries will allow renewable energy technology to overtake traditional fossil-fuel electricity plants.[37]

Cooperation with Europe

Morocco's impressive renewable energy potentials are of special interest to Europe for at least two reasons. First, geographically Morocco is closer to Europe than any other North African country. The short distance means it is easier and cheaper to export electricity and other forms of energy from Morocco to Europe. Second, politically more stable than most of its neighbours to the east (Algeria, Tunisia, Libya and Egypt),

Morocco has enjoyed a higher level of socio-economic stability and has consistently adopted a pro-Western policy.

Against this background one can understand the close cooperation between Morocco on one side and individual European countries and the EU on the other side. This cooperation is not only between governments but also between private banks and institutions on both sides of the Mediterranean. For example, the German Deutsche Gesellschaft fur Internationale Zusammenarbeit (GIZ) advises the Moroccan authority on developing laws and regulations on renewable energy. It also carried out wind measurements at different sites that laid the foundation for investment projects. Another German private firm (Kreditanstalt fur Wiederaufbau) is heavily involved in Morocco's wind industry. Similarly, the French company Areva is likely to play a leading role in initiating nuclear power in the North African country. In 2012 nine energy-efficient pilot projects for buildings in the social housing and tourism sector were launched in Morocco. The projects were funded by the EU in partnership with MDEREE for a total amount of 7.4 million Euros.[38] These nine projects show that it is possible to make significant energy savings with relatively small investments.

In recent decades one of Europe's most serious challenges has been how to respond to the increasing demand for energy without harming the environment. Generally, Europe enjoys one of the highest standards of living, which also means high energy consumption. Meanwhile, with few exceptions, most European countries lack indigenous energy deposits. For decades, Europe has been heavily dependent on imported fossil fuels. With the growing concern about climate change and protecting the environment, the Europeans have sought to diversify their energy mix. Renewable energy has been increasingly seen as the appropriate answer to meet the growing demand with less pollution. Especially solar energy could perfectly balance the triangle of security of supply, competitiveness, and sustainability.

The EU has become a world leader in the promotion of clean energy and in the fight against climate change. Indeed, the development of renewable energy constitutes one of the pillars of the emerging EU's energy strategy, where it has committed to reach a 20 per cent of renewable resources of energy in its total consumption by 2020. Within this context, there has been an increasing interest in 'green electricity' imports from the southern shore of the Mediterranean in order to reach this goal. Morocco is probably the best positioned country within the Southern Mediterranean region to implement the Solar Plan. It already has a relatively significant solar and wind energy installed capacity and, given its proximity to Spain, it has the only relevant and functioning electricity interconnection with the EU in the whole region.

In 2003 the Trans-Mediterranean Renewable Energy Cooperation was formed. This voluntary association calls for an increase of Europe's energy supply and a reduction of its greenhouse gas emissions by campaigning for renewable electric transmission to Europe via high-voltage direct current lines from solar and wind power stations in the deserts of the Middle East and North Africa.[39] One of the earliest steps to institutionalize energy cooperation was the creation of a Priority Action Plan 2008–2013 that contained measures covering a wide area of energy-related subjects.[40] The establishment of the Union for the Mediterranean (UfM) gave a fresh opportunity for cooperation on energy issues and climate change, particularly in the field of solar power. The Mediterranean Solar Plan (MSP) is the most ambitious scheme. It is one of six key initiatives of the UfM, launched in Paris Summit in July 2008. The plan has two complementary targets: developing 20 GW of new renewable energy production capacities, and achieving significant energy savings by 2020. In other words, the MSP addresses both supply and demand. Two years later, experts from the UfM issued a strategy paper, in which they agreed to focus on the following areas:

- Setting up of an adequate legal, regulatory, economic, institutional and organizational environment to enable the development and massive deployment of solar energy and other renewable energy technologies, and to facilitate their exchange or trade
- Examine and promote, in cooperation with European and international financial institutions, the best use of all possibilities to finance investments in renewable energy
- Promote the development of electricity interconnections in order to establish a viable 'green electricity' import and export framework
- Support energy efficiency initiatives and energy savings to realize energy-saving goals by 2020
- Facilitate extensive cooperation on all technology aspects
- Avail all EU carbon mechanisms for the benefits of both Mediterranean sides with the purpose of improving the economics of projects under the MSP
- Continue regular dialogue between stakeholders of the MSP for close coordination and successful implementation.[41]

In March 2012, the European Commission gave its formal approval to the Mediterranean Solar Plan Project Preparation Initiative. This support covers 100 per cent of the costs associated with technical assistance for preparing sustainable energy investments projects in nine Mediterranean partner countries (Algeria, Egypt, Israel, Jordan, Lebanon, Morocco, Syria, Tunisia, and the West Bank and Gaza).[42] Morocco expects the MSP to accompany its policy of renewable energy promotion in order to boost green electricity generation for both internal consumption and export, and transforming renewable energy as a new driver for the country's economic development. Stated differently, by developing green energy, Morocco is providing economic and social solution to itself and to Europe.

The Sahara Wind Project (SWP) is another example of partnership between the two sides. The development of wind energy started as a European success story with Denmark and Germany taking the lead. In order to fully utilize wind power heavy investments have been made in offshore wind potential. Another option is to develop wind power on the other side of the Mediterranean where there are areas with excellent wind conditions and the demand remains limited due to extremely low population density. As the Saharan trade windblown coastline is already accessed by the extension of the European electricity grid, considerable amount of renewable wind generated electricity could be transferred. Initiated in 1993 the Sahara Wind Project seeks to promote the interests of both Morocco and the EU.[43] The site selected for the introductory phase of the project is located in Morocco's main water treatment facility and headquarters of the National Potable Water Office.[44] In facilitating the deployment of renewable energy on a large scale, the SWP enables the transition into a safer, more sustainable energy infrastructure in both continents. The project aims at delivering over 5 GW of wind generated electricity to supply Euro-Mediterranean markets via a High Voltage Direct Current (HVDC) transmission cables.[45]

Desertec

Desertec is one of the largest and most ambitious projects in the world. It was initiated under the auspices of the German Trans-Mediterranean Renewable Energy Cooperation (TREC). RREC was founded in 2003 by the Club of Rome, the Hamburg Climate Protection Foundation, and the National Energy Research Center of Jordan.[46] Desertec concept is based on the proposition (developed by Gerhard Knies, a German particle physicist) that in just six hours, the world's deserts receive more energy from the sun than humans consume in a year.[47] The concept integrates all types or renewable energies including concentrating solar-thermal power in desert regions, wind power in coastal areas, hydro power in mountainous regions, photovoltaics in sunny areas, and biomass and geothermal power where geographic conditions are favourable.

Desertec is a largely German-led initiative. In 2009 an international consortium of companies formed the Desertec Industrial Initiative (Dii) with weighty companies such as E.on, Munich Re, Siemens, Deutsche Bank, and HSH Nordbank signing up as shareholders. By developing the technical and ecological framework in the deserts of the Middle East and North Africa, the initiative seeks to produce sufficient power to meet around 15 per cent of Europe's electricity requirements and a substantial portion of the power needs of the producing countries. More specifically, Desertec seeks to achieve the following objectives:

- Greater energy security in the EU/MENA countries
- Growth and development opportunities for the MENA region as a result of substantial private investment
- Safeguarding the future water supply in the MENA countries by utilizing excess energy in seawater desalination plants
- Reducing carbon dioxide emissions.[48]

The institutions involved in Desertec concluded that the high solar radiation in the deserts of the Middle East and North Africa outweighs the transmission losses between the desert regions and Europe. Thus, it is more economical to install solar thermal power plants in the desert regions than in southern Europe. The electricity network between the two regions will be made up of HVDC transmission cables. These cables can carry electricity generated from renewable resources over long distances with losses of just 3 per cent per 1,000 kilometres.[49] The tentative total cost of the project has been estimated at 400 billion Euros. It is projected that international financial institutions such as the World Bank and major European (particularly German) banks will be the main investors.

The North African deserts were chosen as an ideal location for solar farms as they enjoy strong direct sunlight for much of the year. In addition, the deserts are sparsely populated, making it possible to set up large solar farms. The installation of a 500 MW solar farm near Ouarzazate in Morocco is one of the first stages. The success will help convince both investors and politicians that similar farms could be repeated across the Middle East and North Africa in the coming years. Desertec gained momentum in 2012 shortly after the Fukushima nuclear disaster in Japan. Germany decided to phase out its nuclear power and focus more on renewable energy.

To conclude, from the perspective of the southern Mediterranean countries, such as Morocco, the development of renewable energy in its territory (either through the MSP, Desertec, or something else) has a significant socio-economic potential, particularly for countries such as Morocco with high dependency on imported fossil fuels. Renewable energies can stimulate its economy through the promotion of foreign direct investment, generation of new local energy sources, export of green energy to the EU, creation of employment and fostering of R&D, and transfer of technology. The manufacture and assembly of components in the wind and solar industry can also contribute to the local and national economies.

These potential positive results, however, should not be taken for granted. Still some major issues need to be addressed. These include the harmonization of technical standards, liberalization of energy market, and the development of regulatory regimes, institutions, and laws. The European and Moroccan authorities are aware of these challenges and have already taken major steps to address them.

Conclusion: energy efficiency

In 2010 Morocco's energy mix was comprised of approximately 30 per cent of renewable energy (4 per cent wind and 26 per cent hydro) and 70 per cent of fossil fuels (17 per cent gas, 24 per cent oil, and 28 per cent coal).[50] According to the Moroccan Ministry of Energy, Mines, Water, and Environment, in 2011 the total installed capacity of renewable energy aside from hydropower was approximately 300 MW of concentrating solar-thermal power (CSP), photovoltaics (PV), and wind.[51]

These figures and the analysis above show that Morocco possesses virtually no fossil fuels of its own and therefore must import almost all of its needs from overseas. This heavy dependency on imported oil, natural gas, and coal has had negative impact on the country's finance and environment. Nevertheless, in recent years and in partnership with foreign companies and governments the kingdom has made significant progress in de-regulating its electricity sector and reducing its heavy dependency on imported fossil fuel. The main focus has been tapping the nation's enormous renewable sources and enhancement energy efficiency.[52]

Despite this progress, more work needs to be done in the areas of statutory framework, public awareness of the importance of renewable energy and energy efficiency, availability of the necessary technology, services, and adequate trained labour force. Finally, a more aggressive approach to reform the energy sector and create an attractive investment environment is needed.

Since the early 1990s the Moroccan government has taken some steps to reduce its monopoly over the energy sector and sought to engage the private sector and encourage foreign investment. Seven distribution networks were sold to Mobil, Shell, and Total. The Italian firm Dragofina acquired the Societe Marocaine des Hydrocarbures and Dragon Gas.[53] In the electricity sector, the government has pursued a public–private partnership strategy. In 1963 the National Electricity Office (ONE) was established as a legally and financially autonomous public entity responsible for electricity provision. It operates as a single buyer (from small electricity producers) and owns the entire transmission network, and the majority of the distribution grid. Power shortages and a desire to control public spending have led the Moroccan government to make more use of the private sector to meet the country's power needs. In 1994 the electricity sector opened to private entities and eight years later ONE began restructuring that allowed subsidiaries to be formed. The government has traditionally used subsidies to absorb fluctuations in energy prices and shield consumers from electricity tariff increases, and to expand the distribution network to reach rural populations.[54] Subsidies on electricity prices have been phased out in recent years.

Another step was the enacting of Law no. 13 of 2009. This Renewable Energy Law aims to promote energy production from renewable sources to the local markets and export to neighbouring ones by public or private entities. The law permits electricity to be produced and exported by private entities, but the supply of electricity still must be undertaken through national electricity network.[55]

Finally, Morocco has made substantial progress in encouraging public and private investment in its growing renewable energy sector. Still, there is room for improvement. The experience in other countries suggests that financial incentives (e.g. tax breaks) are needed in the early stages. Equally important, the large subsidies fossil fuels receive need to be addressed. Like in many countries, these subsidies are largely meant to protect the consumers from the fluctuation and ensure that poor people will have access to oil, gas, and coal at affordable prices. In most cases these subsidies fail to achieve their goals and instead encourage overconsumption and waste. In recent years these subsidies have decreased slowly but continue to be considerable and harm the competitiveness of renewable energies.[56]

In addition to seeking to develop renewable resources and other types of energy, the Moroccan authority has been aware of the necessity to address the demand side. In recent years the Renewable Energy Agency (ADEREE) has launched several energy efficiency programs in the construction, industrial, and transport sectors. These energy efficiency programs involve joint ventures between public authorities and international partners. Furthermore, the government requires all its agencies to design an energy efficiency strategy. As such, the construction sector is encouraged to integrate solar resources in building projects (e.g. use of solar water heaters).

This solar water heater program is a United Nations funded initiative. Since its inception in 2002 it has cut carbon emissions by more than a million tons and is projected to reduce around 920,000 tons of greenhouse gas emission a year until 2020.[57] This program and similar ones should help Morocco achieve the goals stated in its National Plan for Renewable Energy and Energy Efficiency – save 12 per cent by 2020 and 15 per cent by 2030 of total energy consumption.[58]

3
Egypt

With more than 90 million people, Egypt is the most populous country in the Middle East and the Arab world. The country's millennium-long history and its cultural dominance have given the Egyptians a strong justification to claim regional leadership. These claims aside, certainly, political and socio-economic developments in Cairo are always echoed in neighbouring countries and impact them positively or negatively. Energy is not an exception.

Over the last century Egypt's energy policies and energy sector have witnessed fundamental changes and growth. In the second decade of the twenty-first century several characteristics of the country's energy sector can be identified:

- Egypt's large population and economic development have boosted demand for energy in all sectors (residential, transportation, and industrial). As a result consumption of oil, natural gas, and electricity has surged in the last few decades.
- For decades production has been largely dominated by fossil fuels (mainly oil and to a less degree natural gas, with coal being used exclusively in industry). Unlike its neighbours to the east (Persian Gulf States) and to the west (Libya and Algeria), Egypt has never been a major oil producer. Cairo holds limited proven oil reserves and until the early 1990s was able to meet its growing demand. In recent years domestic supplies fell short of demand and the gap between production and consumption is growing rapidly.
- The outlook for natural gas is more promising. Over the last few decades Egypt, in partnership with international companies, has made significant explorations and established itself as a major regional producer and exporter.

- Equally important, Egypt enjoys tremendous renewable energy resource potential from solar irradiation in the vast western desert, wind resources along the Gulf of Suez, and hydropower from the Nile River. Despite such potential, the combined share of renewable resources in the energy mix is fairly limited.
- Egypt was one of the first Middle Eastern countries to show interest in nuclear power. Working with international partners and the International Atomic Energy Agency (IAEA), Cairo has developed human and technological nuclear infrastructure. In recent years Egyptian leaders have supported plans to expand nuclear power, particularly in electricity generation.
- Despite long traditions of economic nationalism and state-led economic policy, Egyptian energy sector has been relatively more liberal than some of its neighbours. Generally, Cairo has a long and successful record of partnership and cooperation with major international oil and gas companies. The limited proven reserves have prompted the Egyptian authorities to offer attractive incentives to attract foreign investments. The full utilization of the country's hydrocarbon deposits and renewable potential will depend, to a great extent, on this partnership between the Egyptian authorities and international companies.
- Egypt's key role in regional and global energy markets is based less on its proven reserves and more on its strategic location connecting Africa, Asia, and Europe and controlling two major transit routes (Suez Canal and the Suez-Mediterranean Pipeline 'SUMED'). These two transit routes play an important role in the international trade of oil and natural gas.
- Like most of its neighbours, the Egyptian government heavily subsidizes energy products. These generous subsidies lead to waste, pollution, and inefficiency. In recent years Egyptian leaders have considered different proposals to reduce these subsidies, but concern about popular backlash has restrained any bold initiative.

In short, petroleum and gas exports as well as Suez Canal and SUMED revenues provide a substantial share of Egypt's national income. Comprehensive development and utilization of alternative energy (i.e. nuclear and renewable power) would contribute to Egypt's energy security and economic prosperity.

Oil

The Anglo-Egyptian Oilfields, a joint venture between British Petroleum (BP) and Shell, developed the first oil field in Egypt (discovered in

1869) and brought it into commercial production in 1910.[1] Exploration stopped during World War I. In the following decades production grew and Egypt became a net exporter of crude oil for the first time in 1976.[2] Production peaked at 945 million barrels per day (b/d) in 1993. In 2010 production was approximately 736 million b/d.[3] With ten refineries, Egypt has the largest refining sector on the African continent.[4] In recent years the government has developed plans to expand refining capacity and build more refineries in partnership with foreign companies.

Production comes mainly from six areas: Gulf of Suez, Sinai Peninsula, Nile Delta, Western Desert, Eastern Desert, and the Mediterranean Sea. The Gulf of Suez is by far the most prolific and prospective oil province in the country. In the aftermath of World War II, oil exploration operations were concentrated in the Sinai Peninsula. Similar efforts were made in the 1960s in the Western Desert. Since the beginning of the twenty-first century more attention has been given to the Western and Eastern Deserts. In both areas production costs are lower and lead times are shorter than in other regions. The steady improvement in deep water drilling has encouraged exploration in the offshore Mediterranean Sea. Generally, Egypt lacks super-giant oil fields, like those in the Persian Gulf states. Instead, most of the country's production is derived from relatively small fields.

On the demand side of the energy equation, Egypt's consumption has been on the rise with occasional variation at the sectoral level (transportation, electricity generation, and industrial). In 1990 the country's consumption was approximately 475 million b/d, ten years later it was 552 million b/d, and in 2010 reached 757 million b/d.[5] This surge in demand reflects the rapid population growth and economic expansion. The transportation sector accounts for the largest share of petroleum demand. This is in parallel with surge in car ownership. In 1990 there were 1 million licensed vehicles in Egypt, ten years later the number doubled to 2 million, and by 2010 there were 5.6 million licensed vehicles.[6]

Oil-fired power generation used to account for a large portion of oil consumption. In recent years natural gas has increasingly replaced oil. Furthermore hydropower has traditionally been utilized in electricity generation. Finally industries such as iron, cement, and petrochemicals are major oil consumers.

Petroleum industry is a major economic sector in Egypt. The rise and decline of oil exports have had significant impact on the country's overall economic development. The numerous explorations in the 1960s and early 1970s turned the country into a net oil and petroleum product exporter. Since then most important oil fields have matured and production has declined. In late 2000s Egypt is a marginal exporter of crude and is likely to increasingly depend on foreign supplies to meet its growing demand in the near future. In addition, given the country's large refining capacity, Egypt imports and exports a variety of petroleum products.

Natural gas

In the second half of the twentieth century oil dominated Egypt's energy sector; the slow but steady discoveries of natural gas have changed the country's energy outlook and already made significant contribution to Cairo's national income. Since the early 2000s Egypt has emerged as an important producer, consumer, and exporter of natural gas. Faced with rising demand for energy combined with decline of domestic oil production, the Egyptian authorities turned to gas. The idea was to replace oil by gas in the domestic market and thus save more of the crude and petroleum products for export. Like in many countries, initially the natural gas associated with oil production was flared. There was neither market for it, nor a network to collect and ship the gas from producing fields. This has changed gradually and a vibrant gas industry has been established. Indeed, natural gas is projected to be the primary growth engine for Egypt's energy sector for the foreseeable future.

The first commercial natural gas discovery was made in the Nile Delta in 1967 and production went on stream in 1975. In the following decades, proven reserves, production, and consumption have surged. Proven reserves grew from 0.4 trillion cubic meters (tcm) in 1990 to 1.4 tcm in 2000 to 2.2 tcm in 2010. Production rose from 6.8 billion cubic meters (bcm) in 1990 to 21.0 bcm in 2000 to 61.3 bcm in 2010. The figures for consumption are 6.8 bcm in 1990, 20.0 bcm in 2000, and 45.1 bcm in 2010.[7] Most of these deposits are non-associated gas.

Over 80 per cent of Egypt's natural gas reserves are in the Mediterranean and Nile Delta followed by the Western Desert and Gulf of Suez.[8] Since the early 1980s a partnership between Egyptian authority and foreign companies has persistently and successfully developed these deposits. In 1999, then, Petroleum Minister Sameh Fahmi announced an Integrated Gas Strategy, or a Master Plan, which highlighted the main guidelines for the natural gas industry:

- An export ceiling was fixed at 25 per cent of total production.
- No foreign or domestic gas operator may export gas from Egypt prior to investing in domestic gas market.

- Special incentives were established to encourage foreign and private exploration and production companies to establish marketing franchises.
- Incentives were also established to encourage diversification within the gas industry.
- All businesses within Egypt – whether state-controlled, private, or mixed – were encouraged to convert to natural gas for energy needs.[9]

Ministry of Petroleum and Ministry of Electricity and Energy along with their affiliated holding companies are responsible for drawing and implementing policies for the nation's hydrocarbon industry and the broader energy sector. The Ministry of Petroleum, created in 1973, supervises all oil and natural gas activities. In 2000 the ministry adopted a comprehensive strategy to fully utilize the nation's hydrocarbon deposits. The main elements of this strategy are as follows:

- Develop oil and gas reserves and increase their production
- Satisfy domestic demand for oil, natural gas, and petrochemicals
- Support exports and increase Egypt's income from foreign currency and the state's reserves
- Increase job opportunities and improve human resources
- Develop and restructure the mineral resources sector
- Adopt new technologies.[10]

Ministry of Petroleum pursues these objectives in cooperation with five major state entities – Egyptian General Petroleum Corporation (EGPC), Egyptian Natural Gas Holding Company (EGAS), Egyptian Petrochemicals Holding Company (ECHEM), Ganob El-Wadi Petroleum Holding Company (GANOPE), and Egyptian Mineral Resources Authority (EMRA).

The EGPC was established under Law No. 20 of 1976. It manages the exploration and production operations in its more than 90 affiliated joint venture companies.[11] In August 2001 Minister of Petroleum issued a decree establishing the EGAS as an entity mandated to focus on the natural gas activities, adapting an effective plan to organize the activities of the natural gas resources to add value to the Egyptian economy. Its activities include upstream (exploration, drilling, production) and downstream (processing, transmission, distribution in domestic market, liquefaction, and liquefied natural gas [LNG] marketing). Its main goal is to make Egypt one of the key natural gas players in the region and worldwide.[12] The ECHEM was created in 2002 and assigned the task of managing and marketing Egypt's emerging petrochemical industry.

Its mission is to develop a competitive petrochemical industry based on the use of advanced technology. It seeks to promote investment, facilitate development of new projects, and establish production factories.[13] The GANOPE was established in 2002 to promote development activities specifically in Upper Egypt, which represents over half of the country's total land area.[14] Finally, established in 1896 EMPRA is one of the oldest in the world and the first in Africa and the Middle East. The list of its objectives includes regulating the mining activities, creating and promoting investment opportunities, seeking clean environment, meeting domestic and export demands, and maximizing national reserve of minerals.[15]

In 2009 the Egyptian government created the Regulatory Agency for Gas and Oil Affairs as part of a program to de-regulate the energy sector, promote competition, and encourage private and foreign companies to explore the nation's hydrocarbon resources.[16] The creation of this agency underscored the decades-long strategy Cairo has adopted to encourage foreign investment in the energy sector. Indeed, financial capital, management skills, and advanced technology brought in by foreign companies have always played a vital role in the development of oil and gas industry. Egypt has often offered relatively fiscal terms particularly when oil production started declining in the early 1990s and promising gas discoveries were made. In the mid-1980s a new model for partnership with foreign companies was devised and a clause was inserted retrospectively into some licenses under which profit was divided between the Egyptian authority and foreign investors.[17]

This partnership has persisted and even grown stronger. Indeed international companies are actively participating in almost all operations, both upstream and downstream from exploration and development to shipping.[18] Given Egypt's location in the southern Mediterranean, European companies have generally played a larger role in developing the country's hydrocarbon deposits more than their American and Asian counterparts.

Over the years BP has developed extensive operations in Egypt that include oil, natural gas, and compressed natural gas (CNG). Most of the BP operations are in the Gulf of Suez and the Nile Delta. BP has a long and successful track record in Egypt stretching back several decades. BP business is primarily in exploration and production. In 2012 the company produced about 15 per cent of Egypt's entire production and supplied nearly 40 per cent of the domestic gas demand.[19] Egypt was the first country where Eni expanded its activities abroad. Operations began in 1954 and in the second half of the twenty-first century Eni is one of the major international companies operating in Egypt. More than 470,000 barrels of hydrocarbon per day (corresponding to about 40 per cent of the Egyptian production) are produced from Eni fields, of which approximately 200,000 barrels are Eni's equity.[20] Eni has traditionally been active in oil exploration and development, but in recent years, the giant Italian company has focused on natural gas, particularly in the Mediterranean. Apache is the largest acreage holding in Egypt's Western Desert and in early 2012 was the third largest in the entire country.[21] Shell arrived in Egypt in 1911 to operate two concessions in Gemsa and Hurghada. The upstream business was managed by the Anglo-Egyptian Oil Company (a joint venture between Shell and BP), while Shell Company of Egypt dominated all downstream distribution business in Egypt. In 1964 the Egyptian government nationalized Shell interests, but the company returned and resumed its operations since the mid-1970s.[22]

Other foreign companies operating in Egypt are RWE DEA (Germany), DANA GAS (UAE), British Gas, and Edison (Italy). It is important to point out that the majority of foreign companies operations in Egypt are joint ventures between the international investors and the Egyptian authorities. Joint ventures are arrangements under which a state enterprise and a foreign firm invested stated amounts of capital that can take various forms, including funds, intellectual property, or physical assets and rights to land. The partners share the risk and the reward of the venture in proportion to the capital contributed. Most international companies prefer this investment formula than other arrangements such as production-sharing agreement or technical services contract.

Natural gas has at least two advantages over oil: it is more environmentally friendly and the prices are lower. Thus, Egypt, like many other countries, has sought to encourage replacing oil by gas in the domestic market. Such strategy contributes to restraining pollution and increasing the volume of petroleum available for export. In recent years some of the major gas-consuming industries (i.e., iron and steel, cement, and petrochemical) have increasingly relied more on gas and less on oil to meet their fuel needs.

The Egyptian authorities have also sought to incorporate natural gas into the transportation sector through the use and development of compressed natural gas vehicles and fuelling stations. As of December 2010 Egypt ranks 13th in the world in the number of natural gas vehicles (NGVs). According to the International Association for Natural Gas Vehicles, Egypt has 122,271 NGVs out of a total of approximately

5 million vehicles (2.73 per cent). These NGVs are supported by 119 fuelling stations.[23]

The electricity sector is by far the largest gas consumer (more than half of total gas consumption). Egypt relies heavily on natural gas and oil for electricity generation. Since the late 1990s, following significant gas discoveries and drop in oil production, gas has increasingly replaced oil in the majority of power plants. Gas accounts for over 70 per cent of total electricity generation and the rest being met mostly by oil and hydropower. Growing population and high rate of urbanization have led to a surge in electricity consumption. In 2008 the Egyptian government announced an ambitious plan to satisfy 20 per cent of total electricity generation from renewable energy sources by 2020.

The significant discoveries in the late 1990s enabled Egypt to produce gas than it consumed. This surplus was directed towards export. In the 2000s Egypt established itself as a key gas exporter both in the Mediterranean Sea region and in the international market. Egypt started exporting gas by pipelines since 2003 and as LNG since 2004.

In early 2000s Memorandum of Understanding to build the Arab Gas Pipeline (AGP) was signed between Egypt, Syria, Lebanon, and Jordan. Shortly after the first phase from Arish in Egypt to Aqaba in Jordan with a capacity of 10 billion cubic meters per year was completed and the pipeline started operation in 2003. The second phase expanded the pipeline from Aqaba to El-Rehab close to the borders with Syria. It was completed in April 2007. In the third phase the AGP reached the Jordanian-Syrian borders and started pumping gas to Syria in September 2008. In the same year Syria completed the fourth phase, which runs from the borders with Syria to Homs, close to the borders with Lebanon. In July 2009 an agreement to transport and exchange the Egyptian gas to Lebanon through Syria was signed.[24] There are talks about expanding the AGP to Turkey where it would reach Europe through the Kilis-Turkoglu interconnection and the Turkish national transmission grid. Political instability and security turmoil in the Arab world (particularly in Egypt and Syria) since 2011 have put on hold these ambitious plans. Similarly, the pipeline carrying Egypt's gas to Israel (since 2008) has been attacked several times and the continuation of these exports is uncertain.

Most of Egypt's gas is exported as LNG from two plants one in Damietta west of Port Said and the other in Idku east of Alexandria. Both are joint ventures comprising Egyptian national companies and foreign firms. The complex in Damietta is the oldest. Construction of the facility began in September 2001 and it came on-stream in 2004. It exports LNG to Spanish market via a receiving terminal at Sagunto in Spain. The majority of the gas exported is used to supply gas-fired power stations in Spain. The operating company, SEGAS, is controlled by Union Fenosa Gas in conjunction with Eni and two state-owned Egyptian companies (EGAS and EGPC).[25] The complex has one train and the plan to add another one was suspended due to insufficient supplies.

The other Idku plant became operational in May 2005. Four companies developed the complex: Egyptian Liquefied Natural Gas, Egyptian Company for Operation of Natural Gas Liquefaction Projects, Beheira Natural Gas Liquefaction Company, and Idku Natural Gas Liquefaction Company. It has two trains. Most of this LNG is exported to the United States, Spain, and France. Smaller volume is exported to other European countries, Asia, Canada, and Mexico.

In an effort to coordinate its natural gas production and export policies with other major producers, Egypt has been an active member in the Gas Exporting Countries Forum (GECF). In a meeting held in Tehran, Iran, in May 2001, some major gas-producing nations created the GECF to facilitate cooperation between them. Since then some members such as Iran and Venezuela have sought to transform the GECF into Gas OPEC. Others (i.e., Algeria and Qatar) believe it would take some time for such a transformation to take place. Meanwhile, Russia, the major gas producer and exporter, has taken an ambivalent stand, sending conflicting signals.[26]

Finally, Egypt enjoys a strategic location at the crossroads of three continents (Africa, Asia, and Europe). This strategic location, along with the country's relatively developed human and technical infrastructure, has made Egypt a major international trade route of oil and natural gas. The Suez Canal and SUMED help Egypt to play this role.

The Suez Canal, connecting the Red Sea with the Mediterranean Sea, opened for international navigation in 1869. Egypt assumed control of the Canal in 1956. The Canal was temporarily closed in 1956 and 1967 (until 1975) due to military conflicts. Still, it is widely considered reliable and the shortest link between the East and the West. Over the years, the Suez Canal Authority (SCA) has carried out different projects to increase the capacity of the Canal and ensure safety of navigation. The goal is to enable Very Large Crude Carriers (VLCCs) and Ultra Large Crude Carriers (ULCCs) to cross the Canal. Despite these efforts, only 1 per cent of global crude oil supply is currently shipped through the Canal.[27]

Since the late 2000s three developments have contributed to the decline in the total oil flow through the Canal. First, the global economic slowdown has drastically reduced the demand for oil. In response, OPEC, particularly Persian Gulf members, cut their production. Second, in recent years the international economic system and the global energy markets have witnessed fundamental changes. The rise of China, India, and other Asian giants has transformed them into major oil and gas consumers and importers. As a result, a large (and growing) proportion of oil and gas exports is going East instead of West. Third, piracy and security concerns around the Horn of Africa have led some exporters to travel the extra distance around South Africa to reach Western markets.[28] Despite this reduction in oil flow, the Suez Canal has continued to provide a significant share of foreign currency revenues to the Egyptian government.

On the other hand, LNG transit through the Canal has been on the rise. Natural gas is increasingly becoming the "fuel of choice" globally. In addition, the LNG technology is improving and the costs are falling. Consequently, the share of LNG trade of the total gas trade is rising. Finally, since the early 2000s Qatar has emerged as the world's largest LNG exporter in the world. Qatari gas is shipped to the United States and Europe via the Suez Canal.

Like the Suez Canal, the SUMED connects the Red Sea with the Mediterranean. It was built as an alternative route and a complement to the Canal. The construction started in 1974 and it was inaugurated in February 1977.[29] The SUMED is owned by the Arab Petroleum Pipeline Company, a joint venture between Egypt, Saudi Arabia, Kuwait, the UAE, and Qatar.

To sum up, Egypt's hydrocarbon deposits are not as large as some of its neighbours. The country was a net oil exporter a few decades ago, but the decline in production and the rise in consumption have changed its outlook. In the foreseeable future Cairo will increasingly depend on foreign supplies to meet its oil needs. Natural gas sector looks a little better. Recent discoveries over the last several years have made Egypt more dependent on gas to meet its domestic energy needs as well as a key player in the regional and international scene. Still, Egypt's large and fast-growing population and high rate of urbanization are pushing energy consumption higher. In order to maintain oil and gas production, Egypt needs substantial foreign investment. Traditionally, the country has welcomed foreign investment in its energy sector. Political and security turmoil since 2011 has significantly affected the Cairo's international economic standing. Against this background, Egyptian leaders have sought to utilize alternative energy: nuclear and renewable power.

Nuclear power

Egypt was one of the first countries in the Middle East to express interest in nuclear power. For different reasons the Egyptian efforts to build indigenous human and technical infrastructure lost momentum. Like other Arab countries, Cairo has revived its nuclear plans since the mid-2000s. The toppling of Mubarak regime and the political uncertainty and economic turmoil since 2011 suggest that pursuing nuclear power is not likely to be one of Egypt's priorities in the foreseeable future.

In the spirit of President Eisenhower's Atoms for Peace initiative, Egypt started its own nuclear program in the mid-1950s, when President Gamal Abd al-Nasser created the Atomic Energy Authority and established the Center for Nuclear Research. Another step was the inauguration of a 2 MW reactor at Inshas in the Nile Delta in 1961. The Soviet Union supplied the reactor and controlled the disposal of spent fuel.[30] Upon request, the IAEA provided Egypt with technical assistance and equipment to update the reactor and improve its safety procedures in the early 1980s.

Egypt's second nuclear reactor was provided by Argentina. In September 1992 Cairo signed a contract with Invap, Argentina's leading nuclear organization, to build a 22-MW research reactor in Inshas.[31] The reactor went critical in 1997 and was officially inaugurated in February 1998 by President Mubarak and his Argentinian counterpart, Carlos Menem. It is important to point out that the two reactors have been used for scientific research and peaceful purposes and have been subjected to IAEA safeguards. Both produce radioisotopes for medical, industrial, and agricultural use. Egypt also has medical facilities, accelerators, and other nuclear-related laboratories, including a hot cell laboratory.[32] Finally, Egyptian universities are important partners in research and development activities. Two public universities (Alexandria and Cairo) and a private one (Egyptian-Russian) offer nuclear-related degrees.

There have been several other attempts to expand the country's nuclear infrastructure in collaboration with foreign powers. In the early 1960s, American and German companies placed bids for a 150-MW plant.[33] The deteriorating relations between Egypt and Western powers prompted Western banks to refuse providing the necessary funding, and consequently the project was abandoned. Several factors can explain Egypt's failure to build viable nuclear program before the 1967 war. In addition to economic constraints and lack of technological infrastructure, the Egyptian leadership had never made the strong commitment necessary to carry out such a huge undertaking. Human and financial resources have never been adequately mobilized to achieve this goal. Furthermore, the foreign assistance that Egypt received was not adequate to lay the foundation for a vibrant nuclear program.

In July 1968 Egypt signed the Non-Proliferation Treaty (NPT), but delayed ratifying it as leverage in arms reduction negotiations in the Middle East. After making peace with Israel and adopting a pro-Western foreign policy, the Egyptian leadership decided to ratify the Treaty in February 1981. This step was also taken to enhance the country's chances of receiving foreign technology needed to expand its civilian nuclear program. According to the NPT, state members are eligible for technical assistance. Thus in the second half of the 1970s Egypt negotiated the purchase of nuclear power reactors with the Ford and Carter administrations. The United States, however, was reluctant to supply Egypt with nuclear technology without it ratifying the NPT.[34]

Despite its accession to the NPT, Egypt's access to foreign nuclear technology remained limited, suggesting that other reasons may have been behind the slow progress in the country's nuclear program. In the early 1980s, shortly after the NPT ratification, Egypt negotiated agreements to buy eight nuclear power reactors from Canada, France, Germany, and the United States. These agreements were never implemented. Lack of funding and safety concerns in the aftermath of the 1979 Three Mile Island accident in the United States and the 1986 Chernobyl accident in the Soviet Union were cited as reasons for the suspension.

Like his two predecessors Nasser and Sadat, Hosni Mubarak was never enthusiastic about nuclear power. Under his leadership, Egypt continued its slow and limited cooperation with foreign powers to acquire nuclear technology. Indeed, President Mubarak highlighted two reasons for his reluctant endorsement of nuclear power: heavy debt and alternative sources of energy. In 1992 Mubarak explained why Egypt had not embarked on a nuclear power program by pointing to the high cost of nuclear energy. He estimated that the construction of three or four nuclear plants would cost $18–20 billion. He concluded, "Frankly, I would be leaving a debt for the citizens, a burden on the people. I cannot do this. I do not want to add more burdens than the people can endure."[35] In May 2001 Mubarak stated, "Egypt does not have a need for a nuclear power plant in a country that is full of natural gas reserves."[36]

These two considerations, among others, have convinced the Egyptian leadership not to aggressively pursue nuclear energy. In a surprise announcement in 2006 Gamal Mubarak, the president's son, renewed Egypt's interests in building a nuclear power program. He described this initiative as "a national project that proves that we are strong and capable of doing something fitting of the grandeur of a country that some have begun to doubt."[37] This call to revive the country's efforts to pursue nuclear power was driven by at least four reasons.

1. The steady surge in energy and electricity consumption due to population explosion and economic growth.
2. On the other hand, oil production peaked and started declining. Similarly, almost all hydropower capabilities had been utilized. The significant natural gas deposits have been used to satisfy domestic needs and for export. The heavy domestic demand for gas means that the volume left for export is rapidly shrinking.
3. The safety record of nuclear power has improved. For more than two decades – from Chernobyl, Ukraine (1986) to Fukushima, Japan (2011) – there was no major nuclear accident. In short, nuclear power made significant progress and became more attractive.
4. Meanwhile, Iran, a rival regional power, was rapidly building a sophisticated human and technical infrastructure. Several Arab countries also started developing plans to acquire nuclear power. Egypt, with its strong claims for regional leadership, could not afford to sit on the sideline. Egyptian leaders felt the need to be at forefront in this regional race.

Against this background, President Mubarak announced in October 2007 broad guidelines to revive the country's nuclear program. Particularly the President underscored the need to enhance the regulatory authority of the existing and new energy and nuclear agencies (i.e., the Supreme Council for Peaceful Uses of Nuclear Energy, Nuclear Power Plants Authority, Atomic Power Authority, and Nuclear Material Authority). Working closely with the IAEA, these institutions sought to develop the legal and legislative framework of the nuclear program. The goal was to build four nuclear power plants by 2022.

Within this framework, the Egyptian authority took many significant steps to transform the President's vision into reality, empowering existing laws and institutions and adding new ones. Al-Dabba, a small town on the Mediterranean, was chosen as the site for the first nuclear plant. This choice was based on its low population density and low seismic activity, among other factors. In 2009 the Egyptian Nuclear Power Plant Authority signed a contract with WorleyParsons, an Australian company, for consultancy services to support the delivery of the first nuclear power plant. The contract commences with site and technology selection studies, carries through to design, construction, management, commissioning, and start-up.[38] The project execution is anticipated to take eight years.

In 2006 a Presidential Decree created a State System of Accounting for and Control of Nuclear Materials (SSAC) and the following year (2007) another Presidential Decree activated the role of the Supreme Council for the Peaceful Purposes of Nuclear Energy in policy-making and approving projects, establishing nuclear stations, and ensuring their safety. After consultation with the IAEA, the Nuclear Law was enacted in March 2010 to regulate plant design and operation.[39] The IAEA also agreed to provide technical assistance in developing and training the necessary local expertise.

In the late 2000s, Egyptian officials debated whether the country would seek to enrich uranium itself or importing the fuel from foreign suppliers. In December 2007 Deputy Foreign Minister Ramzy Ezzedine Ramzy stated that Egypt intended to develop an indigenous uranium enrichment capability. Ramzy explained that the NPT does not prohibit nuclear activities, including enrichment, as long as these activities remain peaceful and under the supervision of the IAEA. The Deputy Foreign Minister concluded, 'We reject any move to make Egypt's nuclear program largely dependent on foreign components.'[40] Nabil Fahmy, former ambassador to the United States, echoed the same sentiments, stating that importing enriched uranium from foreign suppliers would be acceptable only if it applied to the entire Middle East region. He asserted that such approach would not be acceptable if it were aimed at 'only limiting the rights of the Arab states.'[41]

The toppling of Mubarak regime in 2011 and the ensuing political and economic uncertainty have raised doubt on the continuation of the nuclear program. It is unclear whether the political will and economic commitment to pursue this goal are still there. Shortly after the regime fell, the facilities in Al-Dabba were attacked and the site was transformed into a housing complex. With improved domestic security, the Egyptian government has renewed its interest in nuclear power. Foreign partners are likely to take a wait-and-see attitude before committing to working with the Egyptian authority to build and operate nuclear power plants.

To sum up, with very high growth in electricity consumption, Egypt has strong motivations to pursue nuclear power. Unlike most of its neighbours, Egypt has more than 50 years of experience of operating research reactor. The main barriers to nuclear power implementation appear to be the 'lack of the necessary funding and low institutional capacity (both in terms of government effectiveness and political stability'.[42]

Renewable energy

Like most of its neighbours, Egypt's energy mix is heavily dominated by fossil fuels. Oil production has already peaked and slowly is declining. Recent discoveries show that the country seems to enjoy substantial reserves, however, given the rising consumption, Cairo has to balance between domestic demand and export. Nuclear power is still in infant stage and there are uncertainties regarding Egypt's financial capability and political stability. With adequate investment and appropriate policies, renewable energy (RE) can contribute to the nation's energy mix and enhance its energy security.

Since 1970s, Ministry of Electricity and Energy has given considerable consideration to RE. In early 1980s, a renewable energy strategy was formulated as part of broad national energy strategy. The goal was for 5 per cent of annual primary energy consumption to be met by RE by 2000.[43] This goal was modified a few times. In 1986 New and Renewable Energy Agency (NREA) was established to act as the national focal point for expanding efforts to develop and introduce renewable energy technologies to Egypt on a commercial scale. The NREA also is responsible for the implementation of energy conservation programs. Specifically, its mandate includes

1. Renewable energy resource assessment
2. Research, development, demonstration, testing and evaluation of the different RE technologies
3. Implementation of RE projects
4. Proposing standard specifications for RE equipment and systems and issuing licensing certificates
5. Providing consultancy services
6. Technology transfer and development of local manufacturing of RE equipment
7. Education, training, and information dissemination.[44]

The NREA pursues these objectives in cooperation with regional and international organizations such as the Egyptian-German Joint Committee for Renewable Energy and Energy Efficiency and Regional Center for Renewable Energy and Energy Efficiency. In February 2008, the Supreme Council of Energy, headed by the Prime Minister, approved a plan to satisfy 20 per cent of the generated electricity by RE by 2020.[45] Egypt has traditionally been using large amounts of renewable energy, especially hydropower, to generate electricity. In recent years other sources such as wind and solar have received more attention and their share in the energy mix is projected to rise.

Hydropower: For millennia the Egyptians have lived and built one of the oldest civilizations around the Nile. Many historians call Egypt the gift of the Nile. Indeed, the Nile has provided Egypt with plenty of water and is the country's main (and almost) only source of fresh water. The Nile also has given Egypt another gift – hydropower. Egypt has a well-developed hydro-electricity sector that has tapped most of the Nile River's hydroelectric potential. About 12 per cent of Egypt's electricity is generated by hydropower.[46] This electricity is generated from four stations – Aswan Dam, the High Dam, Esna Dam, and Naga Hamady barrages. A fifth one – New Asyut Barrage hydropower plant – is schedule for commissioning in 2016.

In recent years Egypt has to deal with a serious challenge to its hydropower sector and, indeed, to all its water needs. Historically, Egypt and, to a less extent, Sudan have drawn most of their water from the Nile and consumed the largest share of the Nile water. Ethiopia announced plans to build a mega hydropower plant in Benshangul Gumuz state, close to the Ethiopian-Sudanese borders.[47] Ethiopia believes that implementing the project is critical in addressing its severe energy shortages and assisting the country in becoming a major electricity exporter. The other upstream states (Burundi, Kenya, Rwanda, Tanzania, and Uganda) signed agreement (the Nile Basin Cooperative Framework) for the equitable sharing of the Nile's water by the eight countries the river crosses. This means that Egypt has lost its veto power to stop any projects along the Nile.[48] How these developments will affect water distribution in the Nile Basin and the Egyptian hydropower sector is yet to be seen.

To sum up, electricity consumption is increasing much faster than capacity expansions. Most of the existing hydropower resources have already been developed. Meanwhile, the share of natural gas in power generation is growing and Egypt has developed plans to further expand electricity capacity by relying on wind and solar.

Wind: Some Egyptian and foreign observers claim that Egypt has the natural resources and the right environment to meet a large proportion of its energy needs by utilizing wind power. Aktham Abou El-Ella, a spokesman for the Ministry of Electricity argues that, '90 percent of the land in Egypt is empty and is suitable for setting up wind turbines.'[49] Stefan Gsanger, Secretary General of the World Wind Energy Association, echoes similar sentiments, 'The wind conditions in Egypt are among the best in the world.'[50]

Since the early 1990s Egypt has sought to develop and utilize its wind potential in cooperation with international partners, particularly Germany, Denmark, Spain, Japan, the European Investment Bank, and the World Bank. One of the first projects was installing a pilot wind farms in Hurghada in the Gulf of El Zayt. In 2001 the Zafarana wind farm along the Gulf of Suez by the Red Sea was constructed. The electricity generated in these two farms is sold to the Egyptian Electricity Transmission Company through the national grid.

In 2003, a detailed wind atlas for the Gulf of Suez coast was published. It concluded that the region has an excellent wind regime and the potential to host several large-scale wind farms. In 2005 the atlas was expanded to cover the entire country and identified more regions along the Nile and in the eastern and western deserts as well as in Sinai with substantial wind potentials.[51] Based on these assessments, the Egyptian authorities have sought to encourage private sector and foreign partners to invest in constructing and expanding wind farms. A large number of companies working on wind power are under the build-own-operate (BOO) model. The combination of this economic policy and Egypt's geographical characteristics has made the country a regional leader in utilizing wind power. Furthermore, the utilization of Egypt's wind potentials will not only diversify its electricity sources and the overall energy mix, but will also provide the government with an opportunity to create badly needed jobs.

Solar: Several millennia ago ancient Egyptians worshipped the sun god Ra. Several thousand of years later, modern Egypt is seeking to utilize the sun to generate heat and electricity. Solar power accounts for a minor share in the country's energy mix, but there are ambitious plans to take advantage of this untapped source. Egypt is one of the most sunlight-rich countries in the world.

Solar is particularly significant for water heating. In recent years several initiatives have been proposed to exploit and develop solar power largely within the framework of the Mediterranean Renewable Energy Program. With the support of the Italian Ministry for Environment and Territory and the United Nations Environment Program, the project sets the stage for large-scale financing of renewable energy projects as essential part of an overall effort to promote regional renewable energy markets.[52]

Egypt's first hybrid solar power plant, Kuraymat, was inaugurated in 2011. The site, south-east of Cairo, was selected in 1997 for several reasons – its uninhabited flat desert location, the high-intensity direct solar radiation, its proximity to an extensive unified power grid, and expanded natural gas pipelines and water sources.[53] Fichtner Solar GmbH, a world leading independent consulting engineering company for solar thermal power plants, is the project developer.[54] The German firm Solar Millennium Group, in particular its technology subsidiary Flagsol GmbH, was the primary supplier of the majority of instruments used in the solar field. Kuraymat collects solar energy through a mirror surface. It features parabolic trough technology, the most proven solar thermal electric technology, integrated with combined cycle power using natural gas as a fuel. Combining the product of natural gas and solar absorption, the hybrid power plant is capable of producing 150 MW of power, a solar share of 20 MW.[55]

Two more projects are planned: the Kom Ombo and the Marsa Alam. Both will receive funding from the World Bank, through its Clean Technology Fund.[56] Egypt also is a contender to a $550 billion project to build solar collectors in North African desert to supply 15 per cent of Europe's energy by 2020. The implementation of these projects, and others, along with Egypt's ability to attract foreign investment will depend on political and economic stability.

Efficiency

Unlike Saudi Arabia, Iran, and the other Gulf States, Egypt lacks abundant energy resources. But, on the other hand, the country is not energy poor. Egypt holds a combination of fossil fuels, nuclear, hydropower, wind, and solar capabilities; some have already been realized and others are in the planning process. Energy security, however, cannot focus only on one side – supply. The demand side needs to be addressed.

Egypt's energy consumption is high and is growing at an alarming rate. This growth is fuelled by large population, economic growth, urbanization, and low level of energy efficiency. Energy consumption per capita and per unit of gross domestic production is very high, particularly in comparison to other countries at a similar stage of economic development.

Egyptian officials emphasize that energy efficiency is an integrated part of the country's broad and comprehensive energy strategy. Egypt has huge potential for energy efficiency gains. There are some initiatives and programs to encourage the industrial and residential sectors to receive financial incentives and technical support for efficiency improvements. In addition, in 2009 the Prime Minister issued decree creating the Energy Efficiency Unit (EEU) as part of the Supreme Energy Council. The EEU membership includes ministries representing the consuming sectors, energy producers, and other related institution particularly those in environment and finance sectors. The EEU mission includes:

- Collecting and reviewing relevant information
- Drafting policy and position papers on priority topics such as renewable energy and subsidy reform
- Help develop an implementation roadmap for energy efficiency programs and initiatives
- Leveraging available resources to meet current needs.[57]

The creation of the EEU is a positive development. However, there is need for a specialized national agency responsible for strategic planning, effective implementation, and assessment of performance. Furthermore, there is a need to mobilize private sector and create the appropriate financial and regulatory environment to attract foreign investment. The lack of coordination between main stakeholders and the absence of mandatory targets, and performance standards such as building codes are hindering effective energy efficiency developments. Most important, energy efficiency will not improve without adequately addressing the subsidy issue.

Since the toppling of the monarchy in 1952 Egypt has moved away from the free-market model and close to state-controlled one. In the 1970s former President Sadat introduced economic reform, but the state has continued to play a dominant role in the economy system. A comprehensive system of public subsidies is a key part of the state-led economy model. For decades energy and food have been sold at much lower price than their market value due to generous state subsidies. The growing population and rising prices have added more constraints on the state budget. In early 2012 subsidies absorbed approximately 28 per cent of Egypt's budget. About two-thirds of that goes toward fuel and energy, with the rest aimed at reducing food prices.[58]

Subsidies are one way to help the low-income segment of the population, but it is not the ideal instrument to alleviate poverty. In addition to their significant cost, subsidies increase state budget deficit and divert public funds from other uses. Subsidies also encourage excessive consumption and waste. Equally important, the prime beneficiaries of across-the-board energy subsidies are wealthy Egyptians, who pay the same low prices for gasoline, cooking fuel, and electricity as do the country's poor, but consume much more in their cars and homes.

Concern about popular backlash and general lack of political will have prevented the government from taking a bold action to cut or reduce subsidies. In recent years there have been some initiatives to re-structure energy subsidies and bring prices close to market value. The government introduces a 'coupon scheme' or 'smart card' to improve the targeting of fuel subsidy. Given the economic and political instability since 2011, it is uncertain if the Egyptian authority will have the will or the capability to phase out subsidies. Efforts to remove energy subsidies should be gradual and transparent.

Conclusion

Egypt faces peculiar energy outlook. It holds 0.3 per cent of the world's proven oil reserves and 1.2 per cent of proven gas reserves.[59] A combination of population growth and economic expansion suggests that these oil and gas deposits are likely to be depleted in a relatively short time. Already oil production has peaked and Egypt is increasingly depended on imported supplies. Meanwhile, the country is trying to balance between its domestic needs of natural gas and maintaining its volume of export. Finally, the hydropower sector seems to have reached its capacity limitations. This means that alternative energy (i.e., nuclear, wind, and solar) are certain to play a crucial role in meeting the country's growing energy needs.

On the demand side, the excessive energy consumption has to be addressed. Simply stated, the country cannot afford this pattern of consumption. True, the Egyptian government will pay a political price for phasing out subsidies, but, there are few, if any, options left. A bold, transparent, and comprehensive strategy to address subsidies is way overdue. Since the toppling of Mubarak regime in 2011, Egypt has witnessed fundamental political and economic changes. The energy sector, like the rest of the country, will be impacted by these developments. At the same time articulating and implementing the 'right' energy strategy will contribute to long-term political and economic stability.

4
Israel

In the late 2000s substantial natural gas discoveries had been made in the Eastern Mediterranean. These gas fields have the potential to change the dynamics of the regional energy markets. These hydrocarbon deposits might promote cooperation between Israel and its neighbours or, more likely, further polarize and deepen the conflict. What is certain, however, is that these discoveries will fundamentally alter the Israeli energy sector.

The decades-long Arab–Israeli conflict has had significant impact on almost all socio-economic and political developments in the region and in the individual states directly involved in the conflict. The evolution of Israel's energy sector reflects both the country's very limited indigenous energy resources and its tense relations with most of its neighbours. First, for decades there was no energy trade between Israel and its neighbours. As is discussed below, for a short time Israel imported oil from Iran and natural gas from Egypt, but these supplies were subject to political changes and came to a halt when the Shah's and Mubarak's regimes were toppled. Second, given this regional mutual hostility and mistrust, Israel has pursued a self-sufficient energy strategy. Electricity is entirely generated domestically and there are no grid connections with neighbouring economies. (The only exception is Israel supplies electricity to the West Bank and Gaza.)

Third, like other developed countries, Israel's energy intensity (ratio of energy consumed to produce a unit of GDP) is fairly low. The service sector with generally low-energy intensity consumption is a major driver of the country's economy and, unlike the Persian Gulf states, petrochemical industry (a high-energy intensity sector) represents a small proportion of the Israeli economy. Fourth, despite favourable low-intensity ranking, pollution is a major challenge. Coal and oil are widely used as primary fuels and in generating electricity. These two fossil fuels are highly associated with carbon dioxide emissions. Fifth, suitable climate conditions and, until recently, lack of indigenous fossil fuel deposits have prompted the Israelis to explore and utilize renewable energy. Indeed, Israeli companies and institutions, in partnership with international counterparts, are taking the lead in developing renewable sources, particularly solar power. [1]

To sum up, with growing population and robust economic growth Israel's energy consumption is on the rise to meet expanding needs for transportation, commercial, and residential sectors. The energy mix is largely dominated by oil and coal. Since its creation in 1948, Israel has been heavily dependent on imported oil and coal supplies. For decades, Jerusalem has sought to reduce this dependency and diversify its energy mix. These efforts were given a boost by the recent discovery of substantial offshore reserves of natural gas.

This chapter examines the Israeli authority's efforts to articulate and implement a comprehensive energy strategy. This strategy addresses both the supply and demand sides of the energy equation. In other words, the Israeli government has sought to secure adequate foreign supplies of fossil fuels and simultaneously explore and develop indigenous deposits. These efforts have been paralleled by similar ones to utilize nuclear and renewable power. On the other hand, laws and regulations have been enacted to promote and enforce energy efficiency. This broad strategy has been pursued in cooperation with the private sector and foreign partners.

Energy outlook: fossil fuels

Since its founding in 1948 the Israeli energy mix has been heavily dominated by fossil fuels (mainly coal and crude oil). Since then Israel has been importing almost all its coal needs and a large proportion of its oil consumption. Coal provides approximately one-third of the country's energy consumption, almost exclusively in electric power generation. All coal supplies are imported from foreign countries (South Africa, Colombia, Australia, Indonesia, Poland, China, and others).[2]

Like coal, Israel's oil production is very limited. Foreign suppliers provide the bulk of the country's oil needs. In recent years Russia, the Caspian Sea, and Africa have emerged as the major exporting regions to Israel.

Efforts in oil exploration have been made as early as 1947 (before the state was officially created in 1948). Petroleum exploration began in

1947 on a surface feature in the Heletz area of central Israel and the first discovery, Heletz-I, was completed in 1955.[3] Since then, a few small wells had been discovered and developed mainly in Kokhav, Brur, Ashdod, and Zuk Tamrur.[4] The 1967 war represented a turning point in Israel's oil sector. After defeating Egypt, Israel occupied the Sinai Peninsula and controlled its oil fields for almost a decade. Thus, by 1971 the Israelis produced 'a total of 43.2 million barrels of oil approximating their country's annual domestic consumption'.[5] Stated differently, in the early 1970s Israel was self-sufficient in oil production due to the lucrative oil fields in Sinai.

This state of self-sufficiency did not last long. In the aftermath of the 1973 war and following lengthy and complicated negotiations with Egypt, Israel withdrew from Sinai and lost control of the oil fields. The Pahlavi regime in Iran became the major oil supplier to Israel in the mid-1970s. This Iranian supplies did not last long either. In 1979 Ayatollah Khomeini led a revolution in Iran and toppled the Shah. Iran's oil exports to Israel came to a halt. Anticipating the fall of the Shah's regime, Israel built up a six-month oil supply reserve in 1978.[6]

In the mid-1980s the Israeli government carried out a comprehensive geological analysis of the entire country. The study provided detailed exploration data and comprehensive maps of all previous petroleum explorations. Since the mid-1990s some modest oil discoveries have been made around the Dead Sea and the offshore Mediterranean Sea.

To sum up, several characteristics of the Israeli oil sector can be identified. First, the decades-long efforts to discover oil at a commercial level have not succeeded. As of now the country's oil production is negligible. Israel remains heavily dependent on foreign oil supplies. Second, Israel has sizeable deposits of oil shale. The World Energy Council estimates that Israel's shale deposits could ultimately yield as many as 250 billion barrels of oil.[7] This non-conventional source of petroleum is 'sedimentary rock containing organic material from which liquid fuel may be extracted at a rate of 15–17 gallons of oil per ton of shale'.[8] Until recently deposits of oil shale have been largely neglected all over the world because it was difficult to separate the fuel from the stone.[9] Usually high oil prices make developing oil shale profitable. Given the lack of indigenous oil deposits and the country's vulnerability to interruption of oil supplies, Israel is trying to utilize its oil shale resources. However, oil shale processing is a water-intensive process and Israel's water scarcity remains a challenge to large-scale shale development.

Third, Israel has two major refineries, one located at Haifa and the other at Ashdod. Their combined production meets all the country's needs of refined oil products. For decades they had been owned and operated by a government entity, Oil Refineries Limited. In order to improve efficiency and increase competition, since the mid-2000s both refineries have been sold to private entities. In July 2006 the Ashdod refinery was sold to Israel's largest fuel retailer, Paz Oil Company. Shares in the Haifa refinery were sold on the Tel Aviv Stock Exchange in the same year. Fourth, the Israeli government continues to play a dominant role in the ownership, operation, and regulation of the energy sector. Since the early 1990s limited measures to privatize the sector have been introduced and foreign companies have been invited to explore and develop the country's energy deposits, particularly natural gas.

Since the mid-1990s Israeli officials have shown great interest in natural gas exploration and development. According to the Ministry of National Infrastructure, Israel is 'actively striving to diversify the sources of energy by the introduction of natural gas as a primary environmentally friendly and cheaper fuel than other forms of energy'.[10] In line with this strategy, in 2007 National Infrastructure Minister Benjamin Ben-Eliezer confirmed that the government was considering the construction of the country's first liquefied natural gas receiving terminal.[11] This idea was never implemented due to lucrative gas discoveries in the Eastern Mediterranean.

The search for natural gas deposits started in the 1950s. In 1958 the Zohar field was discovered followed by Kidod (1960) and Kanna'im (1961). The three fields had very modest reserves. The breakthrough came in the late 1990s and early 2000s when a significant volume of natural gas was discovered offshore, particularly in two fields – Mari and Noa with combined reserves of nearly 1.5 trillion cubic feet (tcf).[12] These reserves have been developed mainly by two entities: The Yam Tetis group, a joint US–Israeli consortium, and British Gas (BG) in partnership with other international companies.

These new and promising discoveries prompted the government to enact the Natural Gas Market Law in 2002. The law seeks to create conditions for the development of the natural gas industry by means of the private sector and to maintain competition between private and foreign companies. It also regulates the activities of the natural gas market.[13] The Natural Gas Authority, established in 2002, is the regulatory body of the industry. It is entrusted with safety maintenance and strategic long-term planning. Its mission is to ensure supply of natural gas, provide licenses, and sets rates and standards of service provision. It also supervises the settlement of any disputes between natural gas market entities and process consumer complaints.[14] The main consumers are the Israeli Electric Company, industries such as Israel Chemicals and Nesher Cement.[15] In addition, natural gas is being considered as a source of inexpensive fuel for water desalination.

More substantial natural gas discoveries have been made offshore Gaza. In November 1999 BG signed an agreement with the late Yasser Arafat, Chairman of the Palestine National Authority, securing Palestine's first exploration license.[16] Shortly after signing the agreement BG announced the discovery of a large gas field, Marine-1, in Palestinian territorial waters (about 15 miles offshore Gaza). These deposits offshore Gaza are believed to be sufficient to meet Palestinians' natural gas needs and additional amounts for export. Since then Israeli and Palestinian officials have negotiated possible natural gas deals. Lack of political trust, however, has complicated such negotiations.

An important conclusion can be drawn from this brief discussion of Israel's energy outlook. The Israeli government, scientists, and private entities have invested substantial resources and effort in diversifying the country's energy mix away from fossil fuels. However, as in the rest of the world, fossil fuels continue to dominate the country's energy sector. At the end of the 2000s oil accounted for almost half of the country's primary energy consumption; coal 35 per cent; and natural gas 11 per cent.[17] This heavy dependence on fossil fuels has made Israel vulnerable to supply interruption and price fluctuation. The recent natural gas discoveries are likely to change these parameters.

For most of Israel's history, energy entrepreneurs searched in vain for oil and natural gas, lamenting the country's lack of energy resources in a region awash in hydrocarbon deposits. Few countries had looked so hard with so little result.[18] Israel's luck started to turn with a modest offshore gas find in 2000. Nine years later (2009) a major offshore gas field was discovered: Tamar. It contains 8 tcf, one of the largest gas finds in the world and, at the time, the largest ever for Israel.[19] The 250-square kilometre structure is about 40 miles (65 kilometres) southeast of Israel's maritime border with Cyprus. It is a joint venture between American and Israeli companies and includes consortium participants Noble Energy, Isramco Negev, Delek Drilling, Avner Oil Exploration, and Dor Gas exploration.[20]

In late 2010, another major gas discovery was made 80 miles (about 130 kilometres) offshore from Haifa and about 29 miles (47 kilometres) southwest of Tamar. Named Leviathan, it is estimated to contain about 16,000 tcf of gas, making it one of the world's largest deepwater gas discoveries of the past decade.

To sum up, since 2009 approximately 24 tcf of natural gas has been discovered offshore from Israel, and the fields are being developed. These abundant reserves in the Tamar and Leviathan fields far exceed Israel's domestic consumption (approximately 200 bcf annually) and could drastically change Israel both domestically and internationally. Israel's electricity sector will likely be able to switch from mainly coal to natural gas. Such a switch will improve Israel's trade balance and reduce carbon dioxide emissions. These new reserves could also transform Israel into a gas exporter. Israeli leaders have already started exploring potential markets. In late 2010, Prime Minister Netanyahu offered to export natural gas to Greece.[21] Given geographical proximity and close political and economic ties, Europe seems an attractive target. Furthermore, European leaders are seeking to diversify their gas suppliers. Of course, if Israel decides to pursue this option, it will have to compete with other gas producers and exporters such as Russia, Norway, and Algeria.

Another important impact of these gas discoveries will be felt in the country's financial sector. Taxes and royalties have become the subject of intense domestic debates since the discoveries of Tamar and Leviathan. The question is, how much should the government take of the profit, and how much should private companies (both national and foreign) be allowed to keep? Thanks to a 1952 law still on the books, Israel offered some of the world's best perks to energy companies, including low royalties and corporate taxes on exploration. The lucrative discoveries prompted Finance Minister Yuval Steinitz to consider changing these terms retroactively, allowing the government to extract better terms on previously assigned leases.[22]

In April 2010, the Minister appointed a committee headed by Hebrew University economist Eytan Sheshinski to examine the issue and propose an updated fiscal system. The Sheshinski Committee released its recommendations in January 2011 in favour of increasing the government's share of oil and gas revenues to between 52 and 62 per cent, up from the current 30 per cent. The tax increase would be phased in over time; fields that start production prior to 2014 would be partially exempt from the tax increase. In March 2011, the Knesset passed legislation approving these recommendations. According to Finance Minister Steinitz, the tax increase will generate one billion shekels annually for 30–40 years, starting in 2015.[23]

Houston-based Noble Energy, Inc. stands to be one of the companies most affected by any changes to Israel's fiscal regime. Owning hefty stakes in Tamar and Leviathan, among other fields, the company is the biggest foreign player in Israel's gas industry. Therefore it does not come as a surprise that Noble, along with private Israeli energy companies, strongly opposed any change in the tax regime.

The impact of and controversy over these lucrative offshore natural gas discoveries have extended beyond Israeli domestic policy to the broader regional context. The fields are located in a region characterized by deep mistrust and overt hostility between Israel and its Arab neighbours. Some of these neighbours are technically in a state of war with each other. Finally, maritime borders have not been established, opening the door for claims and counterclaims. Hoping to replicate Israel's success in finding new energy resources, Cyprus, Lebanon, and Syria have announced plans for holding auctions for licenses to explore for oil and gas. Furthermore, the Palestinians, Syrians, Egyptians, and Lebanese have claimed that these gas fields might be located in their own waters and have accused Israel of stealing their gas deposits. Lebanese leaders, in particular, have been vocal in their opposition to Israeli explorations in the Eastern Mediterranean, claiming that at least part of the gas deposits is located in Lebanon's exclusive economic zone (EEZ). Hezbollah, which represents the Shiites, the largest sectarian group in the country and part of the government, vowed to defend the country's natural resources.

In recent months, both Lebanon and Israel have taken concrete steps to substantiate their claims. The two countries have never defined their maritime border as they are still technically at war. In August 2010, Beirut submitted to the United Nations its version of where the maritime border should be: the exclusive economic zone (EEZ). A few months later (November), it submitted its version of its western border with Cyprus. In mid-2011, the Lebanese government filed a complaint to the United Nations over the recent demarcation of the EEZ at the maritime border between Israel and Cyprus, and Prime Minister Nijab Mitaki created a government commission to defend the country's natural resources in the Mediterranean.

On the other side, Israel has rejected the possibility of indirect talks under a UN umbrella to resolve the dispute. Instead, it has called on Lebanon to begin negotiations on all border issues, not just the maritime border. Furthermore, the Israeli cabinet approved a demarcation of the northern maritime border. The Israeli line gives it more territory than the one that Lebanon submitted to the United Nations.

Rhetoric aside, it is highly unlikely that Israel and its Arab neighbours will go to war over these natural gas deposits. Instead, the Lebanese government appealed to the United Nations, particularly to the UN Interim Force in Lebanon (UNIFIL), to intervene in defining the maritime border. In addition, the Lebanese parliament passed an energy law in August 2010, paving the way for the country's first offshore licensing round.[24] The Palestinians, Syrians, and Egyptians are also likely to intensify their efforts to attract international oil companies to explore for oil and gas in their respective offshore zones. Lebanon has already signed an agreement with a Norwegian company to explore for gas in what it considers its EEZ.

The disagreements over the gas fields in the Eastern Mediterranean are not only between Israel and its Arab neighbours, but also between Israel and Turkey. In the early 1970s, after an alleged coup attempt by supporters of a union with Greece, Turkish troops were sent to Cyprus, and the island has since been divided between a Greek Cypriot south and a Turkish Cypriot north. In December 2010, Israel signed an agreement with Greek Cyprus delineating the sea border between the two nations. Building on this agreement, Cyprus has licensed Noble Energy to explore a block bordering Israeli waters – the same Houston-based company with large stakes in Tamar and Leviathan. Turkey criticized these moves by Israel and Cyprus on the ground that they disregard the rights and jurisdiction of Turkish Cypriots on the island.[25] In recent years Turkish leaders have repeatedly asserted that they would prevent unilateral exploitation by Israel.

Despite Arab and Turkish opposition, Israel (and Cyprus) is moving ahead with the exploration and development of gas fields. It is unlikely that the opposition will be able to restrain the development of these fields. Instead, Arab countries have been preoccupied by the so-called Arab Spring since early 2011. Meanwhile, post-Mubarak Egypt's reservations on exporting gas to Israel have underscored the Jewish state's sense of energy vulnerability and can be seen as an additional reason for Jerusalem's drive to accelerate the development of the gas fields.

In 1979, Egypt became the first Arab country to sign a peace treaty with Israel. Since then, the Egyptian–Israeli relationship has been through several ups and downs, but the two sides have managed to maintain peace. This peace, however, is often described as 'cold'. Generally, Egyptians are not enthusiastic about having close cultural and economic ties with the Israelis. Despite this lack of enthusiasm, the Egyptian government signed a 15-year agreement with Israel in 2005. According to this pact, Egypt agreed to supply 60 bcf of natural gas a year to Israel via an undersea pipeline from the north Egyptian town of El-Arish to the southern Israeli coastal city of Ashkelon beginning in 2008. The Israeli Minister for National Infrastructure, Benjamin Ben-Eliezer, called the agreement 'historic'; and the Egyptian government said that the deal was 'part of a strategy to diversify the nation's gas export markets and to increase gas exports'.[26] The agreement was executed by the East Mediterranean Gas Company, a consortium of the Egyptian General Petroleum Corporation, Merhav of Israel, and Egyptian businessman Hussein Salem.

With the toppling of the Mubarak regime, the pipeline was attacked more than dozen times and the Egyptian authority suspended gas export to Israel. Israel was forced to fire up the Ashkelon and Hadera coal plants to maximum capacity and convert several stations that run on natural gas to heavy fuel oil and diesel. This switch was very expensive in two senses. First, burning coal and fuel oil has adverse environmental impacts because of the higher emissions of sulphur dioxide, nitrogen oxide, and particulate metals. Second, the costs of this switch to alternative fuels were estimated at from $1.5 to $2 million a day.[27] Responding to these supply disruptions and the heightened energy vulnerability, a committee overseeing oil and gas facilities approved an application by Israel's Ministry of National Infrastructure for construction of a buoy that will serve as a terminal for liquefied natural gas (LNG) carriers.[28]

To sum up, natural gas production is projected to substantially increase in the coming years. It is fairly likely that more gas (and oil) will be found, but the scale of discoveries is highly uncertain. These prospects suggest that natural gas will increasingly replace coal and oil in the current energy infrastructure (transportation, residential, and commercial sectors as well as in power generation). This switch will require major changes and investments. It should also be a part of broader strategy to diversify the nation's energy mix.

Nuclear power

Given Israel's limited indigenous energy-proven reserves and subsequently great sense of vulnerability, the government has spared no efforts to utilize other sources of energy. Nuclear power has proved attractive in at least two senses. Unlike fossil fuels, it does not produce greenhouse emission. Additionally, nuclear power means Jerusalem can reduce its dependency on foreign supplies of coal, oil, and natural gas.

Like many other countries, Israel's nuclear program initially needed foreign assistance. This assistance came mainly from two sources: France and to a lesser extent the United States. In the mid-1950s Israel lacked the necessary infrastructure to build its own nuclear program. Cooperation with foreign countries was crucial. France played a prominent role, fulfilling Israel's technological needs in the early stages of building a nuclear infrastructure. The two nations shared commercial and strategic interests. First, both Paris and Jerusalem saw an indigenous nuclear option as a way to maintain a degree of autonomy in domestic and foreign policies in the bipolar environment of the Cold War. The French nuclear industry was young and growing. It needed to establish credentials and gain a reputation on the global scene. Before World War II, France had been a leading research centre in nuclear physics, but it had fallen far behind the United States, the Soviet Union, and Britain.

Equally important, the two nations found a common enemy in Gamal Abd al-Nasser, President of Egypt for most of the 1950s and 1960s. Seeking to establish himself as the leader of pan-Arabism, Nasser supported Palestinian attacks on Israeli and Algerian resistance to the French occupation. This policy alienated both governments. Weakening Nasser became a shared goal for both Paris and Jerusalem. Meanwhile, Israel provided valuable intelligence obtained from its contacts with Sephardic Jews in North Africa in return for the French assistance that would strengthen the Jewish state. A strong Israel that was capable of threatening Nasser would reduce his involvement in Algeria.

Given these common commercial and strategic interests, France provided critical assistance to Israel's nuclear program. At the same time Israel had been an active participant in the French nuclear program from its inception, providing technical expertise. Thus, the Israeli nuclear program can be seen as an extension of this earlier collaboration.[29] The French role was particularly important in the construction of a nuclear reactor at Dimona in the remote Negev desert. Shortly after Britain, France, and Israel withdrew their troops from Suez in 1956, French and Israeli officials reached an agreement on the construction of the nuclear reactor.[30]

Some sources suggest that the cooling circuits and waste facilities were built three times larger than necessary for a 24-MW reactor, an indication that it 'had always been intended to make bomb quantities of plutonium'.[31] Furthermore, Francis Perrin, High Commissioner of the French Atomic Energy Agency from 1951 to 1970, revealed that, while Paris refused to directly provide Tel Aviv with a chemical separation plant, 'it did not interfere with an Israeli request for assistance from a French firm, Saint Gobain Techniques Nouvelles, which built reprocessing facilities for the French nuclear program.'[32] The Office of Science Liaisons was created to provide security and intelligence for the project.[33]

In 1960 the Dimona reactor faced significant hurdles. Shortly after taking office, President Charles de Gaulle reconsidered France's close nuclear cooperation with Israel, demanding that Israel make the project public and submit to international inspection. He also demanded that Israel promise to use the reactor for civilian purposes, not weapons production. Work proceeded, however, and in the mid-1960s the Dimona reactor went critical.

Foreign publications claim that Israel has increased the Dimona reactor's capacity from 23 MW to 50 or even 70 MW, and the general assumption in the international community is that the reactor produces fissile materials (uranium and plutonium), which Israel uses for nuclear weapons.[34] Israel neither confirms nor denies having nuclear weapons under its policy of ambiguity. Furthermore, Israel is not a signatory to the Nuclear Non-Proliferation Treaty (NPT) and refuses to allow International Atomic Energy Agency (IAEA) inspectors to supervise or visit the Dimona reactor.

The Israel Atomic Energy Commission (IAEC) provides little information on the Dimona reactor. According to the IAEC, the reactor was built as part of the national policy to develop the Negev desert. The research conducted at the reactor is designed to broaden basic knowledge in nuclear science and related fields and to provide the foundation for the practical and economic utilization of nuclear energy. The IAEC stresses that periodic authorizations to operate the reactor are granted following extensive safety tests.[35]

The other nuclear reactor is the Center for Nuclear Research at Nahal Soreq, near Tel Aviv. This small 5-MW reactor was provided by the United States in the 1960s as part of the Atoms for Peace Program. The reactor is under IAEA supervision and is visited by international inspectors twice a year. According to the IAEC, construction of the Soreq began in 1958 and it became critical in 1960. The reactor's mission includes radiation control and providing consulting services regarding ionizing and non-ionizing radiation. Furthermore, extensive research is conducted particularly in the field of electro-optics. Like Dimona, periodic authorizations to operate Nahal Soreq are granted following extensive safety tests. But unlike Dimona, Nahal Soreq is operated under IAEA safeguards and is open to group visits.[36]

Despite the fact that Israel has not signed the NPT, it has been a member of the IAEA since the Agency was founded in 1957. Israel participates in the IAEA annual meetings and is a party to several treaties under the auspices of the IAEA. For example, in May 1989 Israel ratified the Convention on Assistance in the Case of Nuclear Accident or Radiological. The Treaty established an international network of cooperation to provide assistance in case of a nuclear malfunction or radiological emergency. Similarly, in January 2002 Israel ratified the Convention on Physical Protection on Nuclear Material. According to the Convention, state parties are obligated to the physical protection of nuclear material within their own borders and during transportation out of the country.[37]

The IAEC supervises all nuclear civilian activities as well as the two nuclear reactors. In June 1952 Israel's first Prime Minister David Ben-Gurion established the IAEC. The Commission is headed by a director general who reports directly to the prime minister. The prime minster is the chairman of the IAEC. The Commission advises the government on all matters related to nuclear research and development, determines policies and priorities, and implements them. The IAEC also represents the state at both national and international institutions.[38]

The IAEC consists of several departments. The Department of Nuclear Engineering is in charge of promoting the development of the nuclear infrastructure and clearing nuclear waste. Other responsibilities include planning and testing advanced reactor technologies and following-up with their implementation, researching desalinization of sea water combined with nuclear energy. The Licensing and Safety Department formulates safety goals, rules, and technical safety criteria, and builds and operates nuclear power stations. It oversees safety in the two nuclear reactors and in the treatment of radioactive waste. In addition, it is in charge of nuclear standardization and overseeing radiation safety.[39]

Despite having a relatively advanced nuclear infrastructure, nuclear power does not contribute to the country's energy mix. The two reactors (Dimona and Nahal Soreq) do not supply energy to the country. Israel continues to rely heavily on fossil fuels with the bulk of supplies imported from overseas. In the late 2000s Israel, like other countries in the Middle East and elsewhere, thought to reduce its dependency on fossil fuels and foreign suppliers. In March 2010 the Infrastructure Minister Uzi Landau announced his government's intention to build civilian nuclear plants, offering to build one as a joint project with Jordan, under French supervision.[40] Jordan rejected the offer saying that such cooperation was premature before a settlement of the Israeli-Palestinian conflict.

The minister explained that Israel wanted a cleaner, more reliable source of energy than the large amounts of coal and oil it imports. Dr. Stelian Ghelberg, head of Ministry of Environmental Protection Noise Abatement and Radiation Safety Department, echoed similar sentiments, 'Israel's energy must be produced by as many types of alternative energies as possible'. He stated that though some people think that a nuclear power plant can explode, 'the truth is far from this. Safe nuclear reactors exist everywhere'.[41] Despite this enthusiasm following the Fukushima nuclear accident, Prime Minister Benjamin Netanyahu said that Israel is reconsidering its plans for a nuclear energy facility in light of what happened in Japan.[42] It is unclear how long this reconsideration will last. What is certain, however, is that Israel will continue its strategy to reduce dependence on imported coal and oil and diversify energy mix.

Renewable energy

Israel has expressed interest in renewable energy since the early years when the state was born. In 1956 David Ben-Gurion, the nation's first Prime Minister stated that the sun was 'the largest and most impressive source of energy in our world and the source of our life for energy, plant, and animal'. Yet, the former Prime Minister added, 'a source so little used by mankind today'.[43] Despite this early interest in renewable energy, the country's energy mix is overwhelmingly dominated by fossil fuels. In 2012 Israel obtained around 43 per cent of its energy needs from coal, 37 per cent from natural gas, and approximately 20 per cent from oil. Renewable energy sources represented only 0.1 per cent of total energy capacity.[44]

In order to promote renewable energy, the authorities have introduced several incentives in recent years. In late 1990s a government committee was established to formulate administrative and legislative means for promoting the use of alternative energies and to offer projects and make recommendations to attract local and foreign investments. In mid 2000s it approved a plan to build two solar power stations in Ashalim with a capacity of 200 to 250 MW.[45] Four ambitious plans were initiated in late 2000s and are likely to have significant impact on the country's energy sector for many years.

First, Oil Alternative Plan: an estimated NIS 2 billion (Israeli shekel) is to be invested annually to finance research and development (R&D) companies involved in creating alternatives for oil. Like many other countries, Israeli transportation sector relies on gasoline and diesel. This sector is responsible for almost half of the greenhouse gases and other forms of pollution. In addition to addressing climate change and pollution, reducing dependency on oil and coal would enhance the balance of payments since the county spends a large proportion of national income on importing its energy needs.

Second, Greenhouse Gas Reduction Plan: In the Copenhagen Climate Change Convention in December 2009 Israel set emission reduction targets – 20 per cent reduction relative to the 'business as usual' scenario in 2020. This plan focuses on energy efficiency, green building, and the potential impact of emission reduction on national industry and export. It seeks to promote the scrapping of refrigerators and air conditioners and increase public awareness and promote behavioural changes in energy consumption.[46]

Third, in 2002 the government set a national goal of 5 per cent renewable energy electricity production to be reached by 2016. In 2009 the goals were changed and an operative governmental decision set a goal of reaching 5 per cent production of electricity from renewable sources by 2014 and 10 per cent by 2020. Those goals were translated by the Ministry of National Infrastructure to an installed capacity goal of 2,760 MW by 2020.

Fourth, in August 2008 the government initiated a 5-year plan for the promotion of the renewable energy. The plan includes encouragement of industrial research. Several ministries agreed to fund research institutions and academic conferences. A new development centre was established in the Negev with the aim of supporting R&D initiatives. The plan also promotes professional training and expanding the labour force and consulting services. It assists in financing international tender applications, international marketing of the Israeli technologies and innovations, and the establishment of a verification centre.[47]

These plans and other programs have been promoted and coordinated by the Renewable Energy Association of Israel (REAI), the nation's main lobbying group for the renewable energy market. The REAI was established in 2009. Its main mission is to lobby different Israeli authorities and ministries in the areas of tax cuts, tax exemptions, facilitation of land availability, and investment grants.[48]

Israel was one of the first countries to adopt solar technologies back in the 1970s. In recent years the country has utilized two solar technologies: solar thermal and photovoltaic (PV). The first includes two main categories: concentrated solar power and solar tower. The latter is a method of generating electrical power by converting solar radiation into direct current electricity using semiconductors that exhibit the photovoltaic effect.

The development of solar energy stems from several geographical and historical characteristics. Israel enjoys abundance of sunny days in combination with huge arid zones. The lack of indigenous deposits of oil and coal and the presence of scores of academic institutions that are actively pursuing innovative research in utilizing energy resources provide the context in which solar industry has made significant progress. Finally, the growth rate of electricity consumption is about 3.5 per cent, higher than most other countries with similar economies. According to the Israel Export and International Cooperation Institute, over 85 per cent of residential buildings use solar collectors.[49]

Most of this solar energy is provided by Arava Power Company. The company is Israel's leading solar developer and a pioneer in mid-size and large-size solar fields using photovoltaic technology. Founded in 2006, Arava Power seeks to supply 10 per cent of Israel's electricity needs through alliances with kibbutzim, Negev Bedouin, and other land owners, especially in the south of the country. In 2009 Siemens, the German conglomerate, partnered with Arava Power and invested $15 million to acquire a 40 per cent stake in the company.[50]

In June 2011 Araba Power inaugurated Ketura Sun, the first commercial solar field in Israel. It consists of 18,500 photovoltaic panels over about 20 acres of land. It will generate 4.95 MW (about 9 million kW-h of electricity per year). Equally important, it will spare the production of some 25,000 metric tons of carbon dioxide.[51] In short, the Ketura Sun is a big leap forward towards diversifying Israel's energy mix and reducing its dependency on fossil fuels. In March 2012 Israel's Public Utility Authority issued licenses for nine solar fields. Since then Arava Power has entered deals to lease land from numerous farms and communities in southern Israel.[52]

Solar power has been the main focus of Israel's renewable energy industry for many years. Harnessing wind energy has never been a priority endeavour compared to projects dealing with generating electricity from the sun. However, in recent years, growing attention has been given to wind power. In the early 1990s Israel installed several wind turbines in the Golan Heights, a strategic and windswept plateau captured from Syria in the 1967 war. This windy plateau is one of the few places in Israel that has enough constant wind velocity to warrant the use of wind power over other forms of renewable energy. However, these old wind turbines, called Mei Golan, used outdated technology that does not meet the expectations and aspirations of Israeli energy innovators.

In September 2010 Israel's Prime Minister Benjamin Netanyahu gave 'national infrastructure project' status to a $400 million wind farm in the Golan Heights, clearing it for fast-track approval by regulators. The new farms will comprise 70 turbines totalling 155 MW.[53] Two companies are leading the wind energy push – Multimatrix and Israel Wind Power. The former is a public company, traded on the Tel Aviv Stock Exchange and the latter is a private company specializing in importing, manufacturing, and construction of renewable energy systems. They formed a partnership with the U.S. energy company AES. Their mission is to introduce the field of wind energy to Israel and leverage the business opportunity inherent in producing clean electricity from wind energy.[54]

Institutions and regulations

Although the Israeli government in principle favours privatization of state-owned companies and free market economy, the energy sector remains largely nationalized and state-regulated, ostensibly for national security reasons.[55] Several ministries and government institutions are in charge of formulating and implementing the nation's energy policy. The Ministry of National Infrastructure's major areas of responsibilities include natural gas, renewable energy, fuel, gas safety, oil and gas exploration, energy conservation, mines and quarries, electricity, water, research and development, oil replacement, and research institutes.[56]

The Ministry of Finance plays a central role in allocating the necessary financial resources and choosing and cooperating with foreign investors. Working with the Prime Minister's Office and the Knesset, among others, the Ministry of Finance articulates and implements tax policies. The Ministry of Interior is responsible for all statutory planning activities. The Ministry of Environmental Protection plays a key role in getting environmental issues on the agenda for energy and climate change.

The Israel Lands Administration is responsible for management of all government-owned land (which comprises most of the state). It is responsible for putting land (including the ones slated as site of power facilities) up for tenders. The Public Utilities Authorities (PUA) is responsible for setting electricity tariffs and standards for the quality of service. It issues licenses to electricity-market participants and advises the Ministry of National Infrastructure on investment projects.[57] The involvement of various authorities occasionally causes delays, confusion, and uncertainty.

When needed the Prime Minister's Office and/or expert committee take the lead in coordinating policies between different ministries and other authorities. For example in late 2011 the Prime Minister's Office headed a committee to set up a definition of the appropriate price for each renewable technology. Representatives from all regulative ministries were involved in this process. Another expert committee provided recommendations on the royalty tax regime for companies engaged in the exploration and commercial development of the recently discovered natural gas and oil deposits in the Eastern Mediterranean. The committee's recommendations would increase the government's share on oil and gas revenues to between 52 and 62 per cent up from 30 per cent.[58]

The process of making energy policy benefits from the contribution of several research institutions and generous public and private funding as well as extensive scientific cooperation with leading foreign universities and think-tanks. The long list includes the following:

- National Solar Energy Center, Ben-Gurion University of Negev
- Gordon Center for Energy Research, Tel Aviv University
- Department of Environmental Sciences and Energy Research, Weizmann Institute of Science
- Grand Technion Energy Program, Technion
- Center for Renewable Energy and Energy Conservation, Arava Institute for Environmental Studies
- Institute for Renewable Energy Policy at IDC Herzliya.

In addition to extensive collaboration between Israeli universities and research institutions on one side and their counterparts in foreign countries on the other side, the Israeli authority signed agreements with foreign governments to promote cooperation and scientific research. One of them is the Bi-national Industrial Research and Development (BIRD). The BIRD Foundation was established by the U.S. and Israeli governments in 1977 to generate mutually beneficial cooperation between the private sectors of the U.S. and Israeli high-tech industries.[59] In 2007 the U.S. Department of Energy's Office of Energy Efficiency and Renewable Energy and the Israeli Ministry of National Infrastructure signed 'BIRD Energy' to consolidate cooperation in the areas of renewable energy, energy efficiency, and industrial research and development.[60]

Conclusion: energy efficiency

Since its establishment in 1948, Israel has been heavily dependent on imported supplies to meet its energy needs. The recent natural gas discoveries in the Eastern Mediterranean could potentially transform the country into an exporter of natural gas. Since the late 2000s the Israeli authority has sought to establish the appropriate regulatory framework. New regimes covering taxes and cooperation between government authorities and private sector have been debated. Equally important, there is a need to reach agreement with neighbouring countries on resolving the maritime demarcation issues. Eventually, Israel's natural gas supplies are poised to substantially increase in the next few years.

However, this is not the whole story. The two sides of the energy equation – supply and demand – need to be addressed simultaneously. Indeed, in the last several years several laws and regulations have been enacted to slow the rise in energy consumption, diversify the energy mix away from fossil fuel, and improve the energy efficiency. This late start to address climate change and pollution can be explained partly by the Jewish state's stance on the Kyoto Protocol, which imposed quotas for greenhouse gas emission reduction in 2004. Israel did ratify the Protocol, but was not an Annex I party to the United Nations Convention on Climate Change and therefore did not have to commit to a specific emission reduction. Since then, Israeli population and economy have continued to grow underscoring the need for energy saving and diversification.

In 2010 the Ministry of Environmental Protection and Ministry of National Infrastructure announced similar plans to address these issues. The Greenhouse Gas Mitigation National Plan and the National Energy Efficiency Program share the same goals and call for specific actions to enhance the country's energy security. The key step is to reduce household energy consumption by replacing household appliances from old devices to more energy-efficient ones (e.g., air conditioners and refrigerator devices). They also seek to raise awareness and education, increase investment in scientific research and technology, and promote green building. Finally, they propose the creation of a special fund to finance these proposals using revenues from a special tax incorporated into consumer tariffs for electricity.

5
Saudi Arabia

All societies require energy services to meet basic human needs (e.g., lighting, heat, mobility, communication) and to serve productive processes (agricultural, industrial, and service sectors). Thus, energy is the lifeblood of human existence and modern civilization. For human development to be sustainable, 'delivery of energy services needs to be secure and have low environmental impacts'.[1] For the last few centuries most of the world's energy supply has come from burning fossil fuels (coal, natural gas, and oil), with the latest providing the single largest contribution in the last several decades. This global energy mix is not likely to change in the foreseeable future. According to a recent report by the ExxonMobil fossil fuels will make up about 80 per cent of total energy consumption in 2040.[2]

Not surprisingly, energy security was primarily associated with the sustainability of oil supply at a low price. Recent geological, economic, and political developments have added more complexity to the pursuit of energy security both at national and international levels. In 2011 the Paris-based International Energy Agency (IEA) introduced a comprehensive definition of energy security. This definition underscores four dimensions: availability (geological), accessibility (geopolitical), affordability (economic), and acceptability (environmental and social).[3] Two conclusions can be drawn from this broad approach. First, there is a strong and growing connection between energy security and environmental concerns (i.e. the transition to a low-carbon economy). Second, both the supply side and the demand side of the energy equation need to be addressed. Very low energy prices in countries like Saudi Arabia lead to over-consumption and waste. Such high consumption rates could eventually eat into export capacity with serious domestic and global ramifications.

With 19.1 per cent of the world's proven oil reserves and 12 per cent of production, Saudi Arabia is a major player in global oil markets.[4] Furthermore, the IEA and the Energy Information Administration (EIA), among others, project that a growing share of the world's demand for oil in the coming decades will be met by Persian Gulf producers, led by Saudi Arabia (and Iraq). Finally, since the late 2000s, the kingdom has maintained the world's largest spare capacity. This large capacity gives Riyadh a leading role in stabilizing global oil markets in time of natural or political crises. The broad economic sanctions against Iran underscore the significance of Saudi Arabia in providing assurances to and maintaining stability in the global energy markets.

This rosy picture, however, does not tell the whole story. True, the kingdom is blessed with abundant hydrocarbon deposits, but these oil and gas deposits are finite. In the last several decades almost all countries in the world, both energy producers and consumers, have invested in diversifying their energy mix, with a special focus on nuclear and renewable power. Another important side of the energy equation is the rapid rising of the Saudi oil consumption. In 2000 the kingdom consumed approximately 1.5 million barrels/day (b/d). Ten years later, it consumed about 2.8 million b/d. In other words, the country's consumption almost doubled within a decade. This trend has significant domestic and international ramifications. Simply stated, more domestic consumption means less oil available for export. Oil revenues provide the backbone of the country's national income. Diminishing revenues would lead to economic stagnation and weaken domestic social and political stability. The soaring petroleum consumption will have global ramifications. Despite all the efforts to diversify energy mix, the world still runs on oil and Saudi Arabia is the leading player. Less Saudi oil export means less supply in the international market, which would push prices higher and deal a heavy blow to the fragile global economy. The global oil market is well-integrated. Under such a scenario, all countries will be impacted.

The good news is that this bleak scenario is not imminent. There is a small window of opportunity where this trend can be reversed. Equally important, the Saudi leaders seem to be aware of the challenge facing their nation and the necessity to take drastic actions to articulate and implement appropriate strategies. Ali al-Naimi, Minister of Oil, argues that using alternative sources of energy that are reliable and sustainable will free more oil for export and 'reduce dependence on hydrocarbon resources and keep them as a source of income for a longer period'.[5] In the last few years the Saudi leaders have taken several initiatives to utilize the nation's largely untapped solar power. Equally important, the authorities have developed plans to build nuclear reactors in coming years.

This chapter will examine Saudi Arabia's efforts to diversify its energy mix away from oil and to utilize alternative energy resources, mainly nuclear and renewable power. Several other countries in the Middle East (and elsewhere) are pursuing similar strategies. The Saudi approach, not different from its neighbours, is to focus on the supply side. In other words, ambitious plans to build nuclear reactors and solar plants are either under consideration or in the early stage of implementation. If these plans are fully implemented, the Saudi alternative energy program would be considered one of the most aggressive in the Middle East. However, this chapter argues, similar efforts need to be made to restrain the surge in domestic consumption. A comprehensive and sound energy security strategy should simultaneously seek to diversify the energy mix and improve efficiency.

In the following section, I examine the drives behind nuclear and renewable initiatives. This is followed by a discussion of the steps that had already been taken to bring these initiatives into fruition. Special references will be made to the efforts to build and develop technical infrastructure, particularly King Abdullah City for Atomic and Renewable Energy (KA-CARE), King Abdullah University of Science and Technology (KAUST), and King Abdullah Petroleum Studies and Research Center (KAPSARC). The following section addresses the need to promote energy efficiency and restrain consumption. The concluding section provides a summary of the main lessons other countries can learn from the Saudi experience.

The drive behind nuclear and renewable energy

Ambitions and aspirations to claim regional (and global) leadership have always played a role in making policy. Several Middle Eastern countries have invested substantial resources and made significant progress in their efforts to diversify their energy mix and utilize both nuclear and renewable power. Some of them such as Israel and Iran are Riyadh's adversaries while others are close allies (i.e. the United Arab Emirates). With the largest economy in the region that holds massive hydrocarbon deposits and as the custodian of Islam's holiest sites, Saudi Arabia claims, with strong justification, a regional leadership status. Watching the race for nuclear and renewable energy, Saudi leaders do not want to be behind other regional powers. Rather, they aspire to be at the forefront leading other countries in utilizing these important sources of energy.

Probably more important than enhancing its leadership's credentials, demographic, economic and geographic forces are driving Riyadh to reduce its heavy reliance on oil in favour of nuclear and solar power. First, Saudi Arabia has one of the fastest population growth rates. For cultural and historical reasons there is very little, if any, family planning. As a result, in the last two decades the population more than doubled. In addition to the fast-growing indigenous population, the kingdom hosts millions of expatriate workers. More residents mean more energy consumers.

Second, the common perception that Saudi Arabia is one of richest countries in the Middle East is not accurate. Still, with $16.267 gross domestic product (GDP) per capita in 2010,[6] the Saudi per capita income is higher than most of the regional powers (with the exception of Israel and the other five Gulf Cooperation Council (GCC) states: Bahrain, Kuwait, Oman, Qatar, and the UAE). In other words, the kingdom has the largest economy in the Middle East, but given its large population, some of its neighbours, with smaller population, enjoy higher per capita income. Generally, affluent people drive bigger cars, go longer distances, live in bigger homes, and use more energy than poor people do.

Third, it is true that Saudi Arabia is blessed with massive hydrocarbon reserves but on the other hand the country suffers from severe water scarcity. There are a small number of lakes and seasonal streams. Meanwhile, the rate of rainfall is one of the lowest in the world. In short, water supply is already exploited beyond its yield.[7] To compensate for this shortage, the kingdom depends mostly on two sources – aquifers and the sea. Aquifers are vast underground reservoirs of water. The kingdom has access to salted water from the Gulf in the east and the Red Sea in the west. Desalination is the process to transform the salted water into potable one. 'With approximately 27 desalination stations producing more than three million cubic meters a day of potable water Saudi Arabia is the world's largest producer of desalinated water'.[8] These plants provide a large and growing share of the water consumed in residential and all other economic sectors. Water desalination is an energy-intensive process and the majority of desalination plants run on oil. In the coming years the kingdom will burn more oil to produce more water.

The combination of these three factors (booming population, developing economy, and water scarcity) has exerted tremendous pressure on the demand for electricity. In recent years the demand for electricity has been rising at an annual rate of 8 per cent and is expected to triple by 2032.[9] According to a recent study in 2010 the electricity consumption per capita in Saudi Arabia was more than three times the world average.[10] In order to meet this surge in electricity demand the Saudi government is investing $80 billion to boost its installed power generation capacity from 46,000 megawatt (MW) in 2010 to 67,000 MW by 2020.[11] Furthermore, the Ministry of Water and Electricity has developed plans to 'cut consumption with efficiency measures, including improvements to power stations and new standards for air conditioning units'.[12]

This surge in the demand for electricity is closely linked to the kingdom's rising oil consumption. More than the other five members in the GCC, oil is the dominant fuel to generate electricity and to power water desalination in Saudi Arabia. This heavy dependence on oil to generate electricity has two major negative impacts: it contributes to air pollution and carbon emissions; and it reduces the amount of crude available for export. This environmental impact is yet to be assessed. Meanwhile, top Saudi officials have already pointed out to the connection between rising domestic oil consumption and shrinking revenues. Khalid al-Falih, the president and chief executive of the national oil company Saudi Aramco stated that domestic energy consumption was expected 'to rise from 3.4 million b/d of oil equivalent in 2009 to about 8.3 m b/d of oil equivalent by 2028 or a growth of 250 percent'.[13] At that rate, Al-Falih concluded that the oil availability for exports is likely to 'decline to less than 7 million b/d by 2028, while the global demand for our oil will continue to rise'.[14] Hashim Yamani, president of KA-CARE echoed similar sentiments, 'Oil exports and economic growth will be constrained if there is no mix of alternative energy'.[15] This projected combination of rising domestic consumption and shrinking export volume is certain to represent a major challenge to internal socio-economic and political stability, particularly given that national income is heavily dependent on oil revenues. In January 2012 Ali al-Naimi, Oil Minister, stated that Saudi Arabia would like to stabilize oil prices at a level around $100 a barrel – higher than the $75 a barrel considered by King Abdullah as a 'fair price' in 2008.[16] This higher price reflects the kingdom's desire to counter the shrinking export volume and the rising public spending. Against this background, developing alternative energy (mainly nuclear and solar) can simultaneously provide fuel to domestic consumption and save oil for export.

Nuclear power

After widespread stagnation in most of the 1980s and 1990s, nuclear power has attracted a growing interest since the early 2000s. Several factors have contributed to this global interest in nuclear power. The threats to oil supplies and fluctuation in prices have raised concerns over 'over-dependence' on fossil fuels. As a carbon-free source, nuclear power can help in slowing the increase in greenhouse gas emissions (primarily from the combustion of fossil fuels). Finally, until the 2011 Fukushima accident, the safety record of nuclear reactors had improved. Despite these advantages, much uncertainty still remains about nuclear power's prospects. One of the main hurdles is the large overlap between civilian and military application. Another is the uncertainty regarding what to do with the nuclear waste. In 2012 there are about 430 commercial nuclear reactors operating in 31 countries and providing approximately 13.5 per cent of the world's electricity.[17]

Saudi Arabia's interest in nuclear power goes back to the late 1970s when it carried out feasibility studies on utilizing nuclear energy to power desalination plants. The kingdom also conducted 'limited experiments to produce radio isotopes'.[18] This early start did not produce any viable nuclear power program. However, the rising domestic demand for energy and the broad and intense interest in nuclear power expressed by numerous states both in the Middle East and worldwide have given momentum to the Saudi nuclear ambition since the mid-2000s. In March 2006 Amr Moussa, then Secretary-General of the Arab League called on Arabs 'to enter into the nuclear club and make use of nuclear energy for peaceful purposes'.[19] Several months later (December 2006), the GCC summit, held in Riyadh, commissioned a study to set up a common GCC nuclear program.[20]

This initiative to coordinate nuclear policy between the six members of the GCC has not made much progress. Instead, every country has since sought to draw its own nuclear power policy. In June 2011 the Saudi government spelled out plans to build 16 nuclear reactors over the next two decades. The goal is to provide 20 per cent of the nation's electricity demand.[21] The 2011 disaster at Japan's Fukushima Daiichi nuclear plant has cast some uncertainty over the future of nuclear power. Some countries have chosen to phase out nuclear reactors while others opted to re-assess the role of nuclear power in their energy mix. Meanwhile the disaster does not seem to have weakened Riyadh's determination to proceed with its nuclear ambition.

Like almost all other nuclear power aspirants, Saudi Arabia needs foreign assistance. Since the late 2000s the Saudi authorities have negotiated nuclear agreements with a number of governments and specialized private companies. In 2011 Riyadh signed three agreements. The first one was signed with France. The pact allows Saudi experts to 'study the French technology options, their financial requirements and implications for developing qualified national human resources'.[22] Argentina's Atomic Energy Commission and Technology firm INVAP has developed a simplified water-reactor design aimed at small-scale electricity generation and water desalination projects.[23] The kingdom expressed interest in this water-reactor design and the two sides signed a deal to boost cooperation. The third agreement was signed with South Korea in November 2011. The two governments agreed to cooperate in the 'design, construction, operation, maintenance, and development of nuclear power plants'.[24] They also agreed to cooperate in research, training, safety, and waste management.

In January 2012 Chinese Premier Wen Jiabao visited Saudi Arabia. One of the outcomes of this high-level visit was the signing of an agreement to promote cooperation in the area of peaceful atomic power. The two sides agreed to work together on the maintenance and development of nuclear power plants and research reactors as well as manufacturing and supplying of nuclear fuel.[25]

The United States and Saudi Arabia have been close allies for decades. The role the US government and American nuclear firms might play in the Saudi nuclear plan however remains uncertain. During George W. Bush's visit to Riyadh in 2008 Secretary of State Condoleezza Rice and Foreign Minister Saud al-Faisal signed a Memorandum of Understanding (MOU) in which they agreed to establish a 'comprehensive framework for cooperation in the development of environmentally sustainable, safe, and secure civilian nuclear energy through a series of complementary agreements'.[26] Two years later, a consortium comprising the U.S. Exelon Corp and the Shaw Group and the Japanese firm Toshiba announced plans to build and operate nuclear reactors in Saudi Arabia.[27] These significant steps, however, remain tentative and little progress has taken place.

In addition to potential substantial commercial profit, the United States has significant strategic interests in promoting cooperation with Saudi Arabia. A Saudi nuclear program built in cooperation with US government and American companies would send a strong message to Iran that full transparency is the 'right' way to pursue a nuclear power program. Washington also would be in a better position to ensure that nuclear materials and know-how would not be transferred to military applications.

Despite these apparent commercial and strategic interests, significant hurdles have prevented real progress between the two sides. One main issue is the uncertainty regarding allowing Riyadh to enrich uranium or reprocess platinum to make nuclear fuel. In its 2008 agreement with the United States, the UAE accepted to voluntarily give up this right. Washington prefers that other nuclear power aspirants follow the UAE model. The Atomic Energy Act Section 123 lists nine criteria that a nuclear pact partner should adhere to (unless the President gives an exemption). One of these provisions bans 'enrichment or reprocessing by the recipient state of transferred nuclear material without prior approval'.[28] Essentially the Atomic Energy Act is a prerequisite for nuclear cooperation between the United States and other nations. Washington and Riyadh have yet to sign a nuclear cooperation agreement. In 2011 lawmakers from both parties sought to amend the Atomic Energy Act to make future nuclear pacts stricter (House Bill 1280).[29] Saudi Arabia has yet to make its stance on this issue known.

The concern by some members in the political establishment in Washington over Riyadh's stance on nuclear proliferation was recently heightened by statements made by Prince Turki al-Faisal, former head of the Saudi intelligence and former ambassador to the UK and the United States. Al-Faisal suggested that if Iran acquired nuclear weapons, Saudi Arabia would do everything to match its neighbour's capability. On the other hand, the kingdom is a partner in a number of international agreements that support the non-proliferation of nuclear weapons. These include the Global Initiative to Combat Terrorism, an international partnership of more than 80 countries that seeks to 'working individually and collectively to implement a set of shared nuclear security principles'.[30] Another agreement is the Proliferation Security Initiative, which aims to 'stop trafficking of weapons of mass destruction, their delivery systems, and related materials to and from states and non-states actors of proliferation concern'.[31] Finally, Saudi Arabia has a Safeguard Agreement with the International Atomic Energy Agency (IAEA). These agreements give the IAEA the power to verify that the signatories live up to their commitments not to use nuclear programs for nuclear weapons purposes. The kingdom, however, has failed, so far, to sign the so-called Additional Protocol, which gives the IAEA additional inspection authority.[32]

To sum up, in pursuing nuclear power Saudi Arabia has already signed cooperation agreements with France, Argentina, South Korea, and China and has negotiated and sought assistance from several other countries including the United States and Russia. This drive to secure foreign partners and the significant steps to build indigenous technological infrastructure suggest that the kingdom is determined to acquire nuclear power.

Renewable energy

Renewable energy is any form of energy from solar, geophysical, or biological sources that is replenished by natural processes at a rate that equals or exceeds its rate of use. It is obtained from the continuing or repetitive flows of energy occurring in the natural environment.[33] Biomass, geothermal heat, hydropower, solar, and wind are all different forms of renewable energy. The use of renewable energy sources is inevitable as fossil fuels are finite. In recent years renewable energy has expanded rapidly and is projected to be the fastest growing fuel (2010–2030).[34]

Renewable energy can potentially provide a variety of environmental, economic, and social benefits. In addition to the reduction of carbon dioxide, and improving health benefits, renewable energy can contribute to the diversification of the energy mix and to the creation of 'green jobs'. Some of the renewable sources can provide electricity to rural and isolated communities cheaper than traditional fossil fuels or nuclear power plants. Some form of renewable resources is available everywhere in the world but their contribution to the energy mix varies by country and by region.

Solar power offers some substantial advantages over other energy sources. Solar-generating facilities are most productive in the middle of the day, when demand for electricity typically is at its peak. Roof-top solar panels can be installed in homes and businesses, reducing the need for centralized power plants and transmission lines. Furthermore, solar power technologies use the least amount of land as compared to other electricity generating technologies. This is because they 'do not require the additional infrastructure for mining, transport, and disposal that nuclear and fossil-fuel powered technologies necessitate'.[35] Saudi Arabia's enormous solar energy potential has long been recognized as an untapped resource.

Saudi Arabia was one of the first countries in the Middle East to utilize solar power. The Saudi Solar Village Project was sponsored by the Saudi Arabian National Center for Science and Technology and the US Department of Energy as part of a joint cooperation agreement signed by the two governments in 1977.[36] This initiative provided electricity to three small villages outside Riyadh. Despite this early achievement, there was no persistent interest in solar power and no similar efforts were made to utilize solar power.[37] Indeed, the kingdom has continued to make little and negligible use of this abundant and clean source of energy.

Sine the mid-2000s, however, Saudi leaders have increasingly shown growing interest in reducing their nation's heavy dependency on petroleum and diversifying the energy mix. In 2008 Ali al-Naimi voiced his nation's ambition, 'One of the research efforts that we are going to undertake is to see how we make Saudi Arabia a centre for solar energy research, and hopefully over the next 30–50 years we will be a major megawatt exporter'.[38] Three years later, Abdullah al-Shehri, governor of the Electricity and Co-generation Regulatory Authority echoed similar sentiment. He stated that the kingdom sees solar power and other non-hydrocarbon sources as 'crucial parts of a plan to boost generating capacity by 50 percent in this decade'.[39] In addition to generating crucially needed electricity, the Saudi authority seeks to create thousands of new jobs in solar farms, assembly plants, and factories that make panels and other equipments.

In 2011 the Saudi authorities took a significant step to translate these ambitious goals into action. The first solar power station, located on Farasan Island, was inaugurated. The plant contains 6,000 solar cells and will produce 864,000 kW.[40] Solar Frontier, the company that built the station, is a joint venture between the government-owned utility Saudi Electricity and the Japanese company Showa Shell Sekiyu. Solar Frontier is currently building another giant solar station in Dhahran called Saudi Aramco North Park that will provide 10 MW of power.[41] Another joint venture between Saudi and Japanese firms is led by the Saline Water Conversion Corporation, which runs dozens of desalination plants in the kingdom, and Hitachi Zosen Corporation. The two sides signed an agreement to carry out research on utilizing solar to power desalination plants.[42]

These Saudi ambitious initiatives to utilize solar power have to overcome several hurdles. One of them is the intermittent character of solar power. In many countries electricity generated by solar plants are supplemented by fossil-fuel-fired stations. The technology has substantially improved in recent years in addressing this reliability challenge. Another hurdle is the need for huge investments to build large-scale solar plants and connect them to the transmission grid. As discussed above, several international companies have agreed to invest billions of dollars to build solar plants in the kingdom.[43] Much more investments are still needed. According to Abdullah al-Shehri, the country will need investments of more than $100 billion over the next decade.[44]

In addition, the authorities need to articulate and approve a regulatory framework for investment in the solar power sector. A third challenge facing the Saudi government is the need to build the necessary human and technological infrastructure to bring into fruition its ambitious plans in both solar power and nuclear energy.

Human and technological infrastructure

In recent years the Saudi leaders have increasingly shown broader understanding of their nation's need to reduce the heavy dependency on fossil fuel and the necessity of establishing the regulatory and educational institutions to build and promote nuclear and solar power. They are not satisfied with importing foreign technology, but rather seek to create indigenous capacity. One of the first steps toward achieving these goals was the establishment of the National Solar Systems (NSS) in 2004. The responsibilities of NSS include:

- Feasibility studies
- Engineering and design services
- Material supply and component sales
- Installation services and supervision
- Commissioning and start-up services
- Operations and maintenance.[45]

The NSS carries out these responsibilities in cooperation with international partners and since its establishment has signed collaboration agreements with several international companies specialized in solar power.

The growing official interest in nuclear and renewable sources since the late 2000s has given momentum to establishing research institutions working on alternative energy. In 2009 the KAUST was founded in Thuwal on the Red Sea. Initially the national oil company Aramco played a major role in funding and managing the KAUST.[46] With one of the richest endowments in the world, the KAUST is a multi-billion dollar research institution with alternative energy at the top of its core research agenda. Researchers at the KAUST carry out their work on nuclear and solar energy in coordination with their counterparts at the KA-CARE.

In April 2010, just six months after the official inauguration of the KAUST, the Saudi government announced the creation of KA-CARE in the capital Riyadh. Article 3 of the Royal Decree highlights the main objective of the KA-CARE, the City 'shall aim at contributing to the sustainable development in the kingdom through utilization of science, research, and industries related to atomic and renewable energy for peaceful purposes'.[47] An official statement adds that developing solar, wind, geothermal resources, along with atomic energy will 'help transition the kingdom to a balanced energy mix and strengthening its ability to meet international demand for oil and make it a world leader in alternative energy'.[48] In order to pursue these ambitious goals, the Royal Decree named Hashim Abdullah Yamani, a Harvard-educated physicist and former minister of commerce and industry to head the KA-CARE.

It is important to point out that the official interest in building and developing technical infrastructure to pursue nuclear and renewable energy has not neglected the oil sector. The authority has recently established KAPSARC as a non-profit institution. The centre's mission is to promote research in energy economics, policy, technology, and environment.[49] Finally in early 2012 Khalid al-Falah, president of Saudi Aramco, announced the launching of a youth enrichment program that will train two million Saudis by 2020.[50]

These programs and institutions, NSS, KAUST, KA-CARE, and KAPSARC, among others, have been tasked to build a strong indigenous technical infrastructure and train Saudi citizens to initiate and support nuclear and renewable energy industry in particular and the energy sector in general. They are well-funded and enjoy support and cooperation from top international counterparts. Still, the learning and training processes take long time and will need sustained official backing for many years to come. Meanwhile, the efforts to expand production and diversify the energy mix should be paralleled by similar attempts to promote and improve efficiency.

The other side of the energy equation: efficiency and conservation

Traditionally, Middle Eastern countries in general, and oil producers in particular, have been lagging behind the rest of the world in terms of energy efficiency (the ratio of energy use to GDP). Indeed, the region's demand for both primary fuels and electricity has been far above the world for many years. The improvement in energy consumption takes place when 'energy inputs are reduced for a given level of service or there are increased or enhanced services for a given amount of energy inputs'.[51] In other words, energy efficiency means the consumption of less energy to produce the same unit of production or service. It improves energy security by increasing the sustainability of energy sources. In addition, by reducing pollution, energy efficiency contributes to cleaner environment.

One approach to enhance Saudi Arabia's energy security is to improve efficiency and conservation. The kingdom is the world's sixth largest oil consumer after the United States, China, Japan, India, and Russia. All these countries have much larger population and more diversified economies than Saudi Arabia. Saudi per capita energy consumption is four times higher than the world average.[52] Saudi citizen consumes a little more than US citizen and around twice as much as Japanese.[53] In recent years economies have grown faster than energy consumption in many countries, but the kingdom is moving in the opposite direction, meaning energy efficiency is getting worse. This poor performance is due to the heavy concentration on energy-intensive industries and the largely energy-intensive lifestyles in residential and transportation sectors. Furthermore, this lifestyle is heavily supported by very low fuel prices and electricity. These subsidies constitute a significant share of government spending.

Like their counterparts in other oil-producing countries, Saudi citizens enjoy generous social and economic benefits, including heavily subsidized petroleum products. A subsidy exists when there is a 'financial contribution' by a government or public body that confers a 'benefit'.[54] It is the 'difference between the market price and the real opportunity cost of the commodity'.[55] Indeed, the prices of fuels, electricity, and desalinated water are just a fraction of their market values. Proponents of subsidies argue that they shield the population against price fluctuations, ensure that the poor have access to electricity and fuel, and buy domestic political support.

However, since these government subsidies do not distinguish between the poor and rich, the greatest benefit goes to those who consume the most energy, namely wealthier firms and households. Thus, some argue that subsidies are not considered an appropriate way to distribute public wealth. Furthermore, they encourage over-consumption, waste of precious natural resources, and contribute to pollution. Thus, lifting or reducing subsidies have been under consideration in recent years. The experience in many countries shows that this approach is likely to be strongly resisted by large segments of the population who have grown accustomed to low energy prices. Political upheavals in the Arab world since early 2011 have further complicated this debate concerning slashing subsidies. Saudi Arabia, and the other Gulf states, had opted for increase in public benefits and spending. Thus, given the political environment, heavy government subsidies and low energy prices are likely to endure in the foreseeable future.

This, however, does not mean that the Saudi authorities have given up on trying to reduce consumption. In 2003 the government established the National Energy Efficiency Program to 'assist the energy sector to meet the rapidly growing power and energy demand through efficient and rational consumption patterns'.[56] The program supported energy auditing and regulations, information exchange, and promoted efficient technologies. In November 2010 the cabinet replaced the program with an independent Center for Energy Efficiency. The centre is housed at King Abd al-Aziz City for Science and Technology (KACST) and chaired by its president Dr. Mohammed Ibrahim al-Suwaiyel. Representatives from the private sector and several government entities are members in the centre. The centre's goals are similar to those of the National Energy Efficiency Program. It seeks to promote energy conservation and study the appropriate methods that suit the kingdom's environment and culture.[57]

Conclusion: the way forward

Most analysts agree that all countries – consumers and producers – need to diversify their energy mix to enhance their security. Each source of energy has its own limitations, advantages, and disadvantages. Saudi Arabia has abundant oil and natural gas deposits, but these fossil fuels are finite and burning them has significant negative impact on the environment. The kingdom, like other oil producers, is seeking to reduce its overwhelming dependence on oil and petroleum products. Nuclear and renewable powers are not panacea, but developing other sources of energy will make the country less vulnerable to the fluctuation in prices and to rising pollution. While nuclear power is broadly perceived as more attractive than it was in the 1980s and 1990s, there are unresolved issues. These include the fuel cycle, waste management, and association with nuclear weapons proliferation. On the other hand, the prices of renewable technology have been falling in recent years, but still renewable sources have hard time competing with heavily subsidized fossil fuels. This dilemma needs to be addressed as part of a comprehensive national energy strategy.

Diversification of the energy mix will also enable Riyadh to keep producing and exporting oil for longer time. A recent report by the British Petroleum projects that the Middle East exports (currently supply

22% of global demand) will rise to 25 per cent by 2030.[58] Most of these additional supplies will come from Saudi Arabia (and Iraq). Failure to diversify the kingdom's energy mix and to restrain souring domestic consumption would weigh on Riyadh's ability to deliver the projected supplies to global oil markets.

In closing three recommendations can be made. First, in the short-term Saudi Arabia is not likely to confront energy crisis. However, doing nothing or 'business as usual' would bring the crisis faster and makes it deeper. In addition to investing in and developing alternative energy the kingdom needs to implement reform and slash subsidies. The large gap between domestic fuel prices and their market value and production cost needs to be closed. With massive financial assets from oil revenues the government is in a good position to implement these unpopular measures sooner rather than later.

Second, the Saudi government needs to adopt a comprehensive approach to introduce and implement energy reform. This means increasing public awareness of the pros and cons of alternative energy, lifting subsidies, and curbing demand. Broad and sustained educational and media campaigns are needed to reach all segments of the population. Academic institutions, Islamic organizations, and the private sector should play a role in educating the public and promoting new values and attitudes towards energy consumption.

Third, the success or failure of a Saudi comprehensive energy strategy will have huge international ramifications. With its large idle production capacity and its leading role in Organization of Petroleum Exporting Countries (OPEC) and other energy forums, the kingdom is considered the world's energy superpower, or central bank. A sound and balanced Saudi energy outlook will not only contribute to domestic socio-economic and political stability but also will enhance the stability and predictability of global energy markets. The US and European governments need to play the role of a major partner in articulating and implementing a comprehensive and balanced Saudi energy strategy.

6
The United Arab Emirates

Energy security has often been associated with the non-interruption of oil supplies. This association fails to capture the growing significance of other sources of energy and the fact that energy security is a major concern for both consumers and producers. A recent study by the International Energy Agency (IEA) highlights four dimensions of the concept: availability (geological), accessibility (geopolitical), affordability (economic), and acceptability (environmental and social).[1]

This chapter seeks to examine the United Arab Emirates' (UAE) efforts to enhance its energy security. The UAE is a federation formed in 1971 and comprises seven emirates – Abu Dhabi, Ajman, Dubai, Fujairah, Umm al-Qaywayn, Ra's al-Khaymah, and Ash Shariqah. Abu Dhabi is, by far, the largest and richest emirate. Since its creation, the UAE has emerged as a major economic power-house in the Gulf region and the broader Middle East. With $358.1 billion nominal gross domestic production (GDP) in 2011[2], the Emirates has the third largest economy in the Middle East (after Saudi Arabia and Iran) and one of the highest per capita income in the world.

This very bright economic outlook is due, almost exclusively, to the Emirates' massive oil reserves. The country's proven reserves are estimated at 97.8 billion barrels (7.1 per cent of world's total), the sixth largest in the world after Saudi Arabia, Venezuela, Iran, Iraq, and Kuwait.[3] The UAE is a member in the Organization of Petroleum Exporting Countries (OPEC) and has been a major and reliable oil producer and exporter.

Despite these massive hydrocarbon deposits, in the last several years the UAE has launched several initiatives to diversify its energy mix and reduce its dependence on oil. Indeed, Abu Dhabi has taken more concrete steps to utilize nuclear power and renewable energy than most of its neighbours. Most Arab countries have recently expressed interests in nuclear and renewable sources, but few have taken any action to translate these interests into executable plans. The UAE undoubtedly is at the forefront.

The following sections seek to address this apparent contradiction – why a country blessed with such massive oil deposits has allocated substantial resources to develop nuclear power and solar? The following section briefly discusses the country's oil and natural gas sectors. After analysing the drives toward nuclear and renewable sources I examine the Emirates' efforts to build nuclear reactors and to use its abundant sunshine. Finally, I argue that these plans to reduce dependence on oil and add other sources of energy should be supplemented by efforts to promote efficiency. High and fast-growing energy consumption in the UAE and other Middle Eastern countries needs to be adequately addressed.

The hydrocarbon sector

In the late 1920s, when the world market for the Gulf's high-quality pearls was destroyed mainly due to the Japanese invention of cultured pearls and the global depression, the already poor emirates fell upon even harder times. Their fortune turned just before World War II when the prospects that massive oil deposits might be trapped underneath their desert emerged. Around this time, oil deposits had already been discovered in neighbouring Iran, Iraq, Bahrain, and Saudi Arabia.

In the early 1930s, the first geological survey was conducted in Abu Dhabi and carried out by a consortium of several international oil companies including British Petroleum (BP), Shell, Total, Exxon, Mobil, and Petroleum Development Company. World War II put all exploration and development activities on hold. Shortly after the end of the war the first well was drilled in Ra's Sadr. The first commercial oil discovery was made at Bab in 1960. Since then several oil fields, both onshore and offshore, have been discovered and developed. In the second decade of the twenty-first century, the list of producing oil fields includes Abu al-Bukhoosh, Asab, Bab, Bu-Hasa, Bunduq, Falah, Fateh, Lower Zakum, Mubarraz, Qusahwira, Rashid, Sahil, Satah, Shah, Umm al-Salkh, Umm Shaif, and Upper Zakum.[4] Much of the production comes from the Zakum oil structure – a collection of oil fields which together are considered one of the largest oil zones in the world. Most of these fields are onshore and offshore. Abu Dhabi, which holds the bulk of the nation's proven oil reserves, provides most of its production.

These oil discoveries have transformed the UAE into one of the richest countries in the Middle East and the entire world. This massive wealth is based on oil exports and revenues. These oil revenues provide the lion's share of the national income and public expenditures. Unlike other oil-producing countries in the Persian Gulf and the broader Middle East, the UAE has always adopted a friendlier and more benign attitude towards foreign investment. Oil companies from France, Japan, the United Kingdom, and the United States, among others, own close to half of the energy sector in the UAE. International Oil Companies (IOCs) have forged a close partnership with the UAE authority from the early 1950s and have enjoyed more hospitable investment environment than in other countries. Major IOCs such as BP, ExxonMobil, Partex, Petrofac, Shell, and Total have played a key role in developing the Emirates' oil industry. Furthermore, the UAE largely escaped the wave of nationalization which swept most of oil-producing countries in the 1970s. Instead of nationalizing foreign companies and confiscating their assets, the UAE created its own national companies, which worked side-by-side with their foreign counterparts.

The Abu Dhabi National Oil Company (ADNOC) was established in 1971 to operate in all areas of oil and gas industry, including exploration, development, production, support services, refineries, processing and petrochemicals, shipping and transportation, and distribution. These operations are carried out by more than a dozen of subsidiary companies.[5] The ADNOC operations are guided by the Supreme Petroleum Council (SPC). Established by Law no (1) of 1988, the SPC is the highest authority responsible for all petroleum affairs. It is in charge of the formulation and implementation of all policies and objectives in all sectors of the petroleum industry. As such, it is chaired by Sheikh Khalifa Bin Zayed al-Nahyan, President of the UAE and Ruler of Abu Dhabi. The other nine members are all from the royal families.[6]

One of the main missions of the ADNOC is to educate and train indigenous Emirates labor force. In addition to sponsoring scholarship programmes in major universities in foreign countries, the ADNOC has founded a number of educational and research institutions in the UAE. Established in 1979, the ADNOC Technical Institute is one of the oldest. It offers training programs that meet the needs of ADNOC and its partners for skilled national manpower and identifies young UAE nationals with a potential for upward mobility.[7] The Petroleum Institute (PI) is another leading institute. Created in 2001, it offers graduate and undergraduate degrees in chemical, electrical, mechanical and petroleum engineering, metallurgical and polymer science and engineering and petroleum geosciences.[8]

The Abu Dhabi Company for Onshore Operations (ADCO) was incorporated under Law No 14 for 1978. It is responsible for the operation of oil fields onshore and in shallow coastal water of Abu Dhabi. Its mission includes developing technical capability, evaluating, determining, and executing optimum reservoir development, and adopting appropriate technologies and strategies.[9] A large share of the Emirates' oil production comes from offshore fields. The major national company in charge of managing these offshore deposits is the Abu Dhabi Marine Operating Company (ADMA-OPCO). In the early days of oil exploration, operation surveys were carried out by oil companies. These surveys found huge commercial deposits in the seabed of two oil-bearing structures – Umm Shaif and Zakum. For decades oil fields from these two areas have provided a substantial proportion of the UAE oil production.[10]

Another subsidiary entity of the ADNOC is the Abu Dhabi National Tanker Company (ADNATCO). Established in 1975 for the transportation of petroleum products, the ADNATCO owns and operates a fleet of oil tankers and other vessels.[11]

Given the key role oil production and revenues play in the UAE's economic prosperity and political stability, the government is determined to maintain the high level of capacity and to further boost production. In order to carry out these ambitious objectives the UAE has to allocate substantial financial resources in upstream and downstream oil sector. It is estimated that from 2004 to 2011 Abu Dhabi invested roughly $7 billion and anticipates investing some $43 billion by 2020.[12] Finally, given the volatility of the Persian Gulf region and the threats to oil shipments via the Strait of Hormuz, the UAE built a pipeline that bypass the Strait. Linking the oil fields near Abu Dhabi with the port of Fujairah in the Indian Ocean, the pipeline became operational on 15 July 2012. This $3.5 billion pipeline has a capacity of 1.5 million barrels/day, or about 65 per cent of the country's exports.[13]

To conclude, the UAE continues to be heavily dependent on oil as the dominant fuel in its energy mix. This dependence underscores the nation's vulnerability. In the last several years Abu Dhabi has been able to maintain the level of its proven reserves primarily due to enhanced oil recovery (EOR) technologies that are used to increase extraction rates from its mature oil fields. High oil prices since the late 1990s have made extraction from mature fields profitable. In recent years few new discoveries have been made.[14]

While oil has fuelled rapid economic prosperity in the UAE, natural gas has recently played a key role in meeting the nation's large and growing demand for energy. Natural gas is largely used in power generation both for residential sector and in water desalination. Gas is also injected in old and mature oil field to maintain its level of production. Finally, gas is used to feed the fast-growing petrochemical industry. The majority of the UAE's gas reserves are located onshore or offshore Abu Dhabi with small amounts found in Sharjah, Dubai, and Ras al-Khaimah. Indeed, Khuff reservoir under the Abu al-Bukhoosh and Umm Shaif oil fields is considered one of the largest in the world.

In order to utilize the nation's natural gas deposits the authority founded the Abu Dhabi Gas Industries (GASCO) in 1978 as a joint venture between the Abu Dhabi National Oil Company (ADNOC), Shell, Total, and Partex. It is responsible for onshore gas processing and transportation of gas and liquid products.[15] One of its major projects is the Integrated Gas Development (IGD). This adventure seeks to link and add new onshore and offshore gas processing facilities in Abu Dhabi. The goal is to make more gas available to meet the growing domestic consumption and to supply feedstock to the petrochemical industry.[16]

The National Gas Shipping Company (NGSCO), formed in 1993, is in charge of transporting the Emirates' liquefied natural gas (LNG). Initially, the LNG fleet was managed by a third party, but since the end of 2007 all vessels are fully managed by the NGSCO. The company operates a fleet of approximately 10 LNG carriers.[17]

The UAE holds the world's seventh largest proven gas reserves (after Russia, Iran, Qatar, Saudi Arabia, Turkmenistan, and the United States, respectively). Its total reserves are about 3.2 per cent of the world's total; however, its share of world's gas production was only 1.6 per cent in 2010.[18] This large gap between proven reserves and actual production is due to the fact that the nation's gas is sour (low quality) and expensive to produce. Meanwhile consumption is surging due to economic expansion and high population growth rate. According to the first comprehensive report on the country's economy carried out by the National Human Resources Development and Employment Authority, the UAE population doubles approximately every 8.7 years as opposed to every 55 years for the world's population.[19] This very high rate is largely due to the country's huge expatriate communities.

As a result, since 2007 domestic consumption has outstripped production and the nation has become increasingly dependent on foreign supplies (both piped and transported LNG), particularly from neighbouring Qatar. In 1999 the United Arab Emirates Offsets Group (UOG) proposed building an undersea pipeline from Qatar to Abu Dhabi. From there the gas would be piped overland, first to Jebel Ali and then on to Oman, with long-term plans to extend it to

India and Pakistan. This project was named the Dolphin Initiative. This proposal was seen as serving the two neighbours' interests. Qatar, with its massive production, needed market for its gas, and the UAE, with its growing consumption, needed secure supplies. In addition, the scheme would contribute to on-going efforts to promote economic and political integration among Gulf states. The two sides signed an agreement in September 2004. First gas delivery was originally due in late 2006 but because of a territorial dispute between Qatar and Saudi Arabia the inauguration was delayed. After extensive mediation and improvement in the Qatari-Saudi relations Riyadh allowed the pipeline to operate in February 2008.[20]

The UAE also tried to import natural gas from Iran. In 2001 Crescent Petroleum Company, a privately owned oil and gas firm, signed an agreement with the National Iranian Oil Company (NIOC), under which Iran would export 17 million cubic meters (600 million cubic feet) of gas daily to the UAE for 25 years starting in 2005. The gas was supposed to be shipped via a pipeline linking Iranian facilities from Salman field in the Persian Gulf to receiving facilities on the Sharjah coast. However, as gas prices rallied in the second half of the 2000s, the two parties could not agree on what is the 'fair price'. In August 2010 the NIOC said the 25-year contract had been nullified.[21]

To sum up, the UAE natural gas imports and production are likely to increase. However, this increase will be offset by a significant growth in gas demand.[22] Stated differently, the large and growing gap between production and consumption is projected to grow wider in the coming years. Other energy sources are needed.

Alternative energy: why?

The motives behind the UAE's interest in and plan to develop nuclear power and renewable energy are not different from those of other nations. These are basically to meet the growing energy needs of its population and to diversify its energy mix. The UAE, like other Gulf states, has one of the world's fastest population growth rates. This population, particularly the nationals, enjoys one of the highest per capita incomes in the world. Rapidly growing population and widespread affluence add pressure on electricity demand. Furthermore, the Gulf region suffers from chronic shortage of indigenous water supplies, with very little rain. This means that Gulf governments depend heavily on desalination to provide their people with adequate water supplies. Desalination process requires substantial amount of fuel and electricity. The combination of these factors – growing population, rising affluence, and water shortage – has led to a surge in oil consumption.

In 2000 the UAE consumed 396,000 b/d. Ten years later, 2010, the consumption level reached 682,000 b/d.[23] This surge in consumption was not matched by rise in production. In other words, in a 'business as usual' scenario, with growing amount of oil consumed domestically the UAE's volume of crude export would decline. Such a scenario would not only have negative socio-economic and political ramifications on the UAE, but will also contribute to a shortage of available oil supplies in the global markets.

Substantial proportion of this domestic oil consumption is used to generate electricity. The UAE officials understand that oil is an exhaustible fuel and that the large volume of domestic oil consumption means lost economic opportunity costs (loss of export revenues). In addition, high domestic consumption has made the country one of the world's largest carbon emission per capita. Other fossil fuels cannot mitigate these negative impacts. The nation consumes more natural gas than it produces. Stated differently, the UAE is a net gas importer. A large proportion of gas is injected into mature oil fields to maintain current level of production. The Emirates lacks adequate coal deposits and burning coal would further add to environmental concerns.

In 2008 Abdullah bin Zaed Al-Nahyan, then the UAE Minister of Foreign Affairs, summed up his country's drive to build nuclear reactors. Nuclear energy, the minister stated, 'represents a proven economically competitive and environmentally promising method for producing electricity'.[24] These are the same reasons behind the Emirates' heavy investments in renewable energy.

Nuclear power

The six Gulf Cooperation Council (GCC) states – Bahrain, Kuwait, Oman, Qatar, Saudi Arabia, and the UAE – have expressed interest in nuclear power as early as 2006 when they ordered a 'GCC-wide study be conducted to formulate a joint program in the field of nuclear technology for peaceful purposes in keeping with international standards and regulations'.[25] Since then the UAE has taken several steps to launch its own nuclear program, including establishing the necessary domestic educational and regulatory institutions and signing agreements with international partners. In April 2008 the government released a policy paper titled, 'Policy of the United Arab Emirates on the Evaluation and Potential Development of Peaceful Nuclear Energy'. In this important document Abu Dhabi not only outlined the reasons for pursuing nuclear power, but also provided the main principles that guide its nuclear energy policy. These include the following:

- Commitment to complete operational transparency
- Commitment to pursuing the highest standards of non-proliferation
- Commitment to the highest standards of safety and security
- Commitment to conform to the International Atomic Energy Agency's (IAEA) standards
- Commitment to develop partnerships with international governments and firms
- Commitment to long-term sustainability.[26]

Since then Emirates officials have sought to put these principles into action. The underlying objective is to build a highly advanced nuclear infrastructure composed of several reactors, in partnership with developed nuclear powers and international corporations. This nuclear infrastructure should be run, managed, and regulated by Emirates nationals, regulations, and laws and should be in line with the highest safety standards and confirm to the global non-proliferation norms and rules.

In 2009 the UAE government enacted new law regarding its newly launched nuclear programme. The law created the nation's regulatory commission – the Federal Authority for Nuclear Regulation (FANR). The FANR's mission is to 'protect the public, workers and the environment by conducting nuclear regulatory programs in safety, security, radiation protection and safeguards'.[27] A few months later, the UAE President Sheikh Khalifa Ben Zayed Al-Nahyan signed a new law establishing the Emirates Nuclear Energy Corporation (ENEC). The ENEC is in charge of the development, ownership, and operation of nuclear reactors. Specifically, its responsibilities include the following:

- Overseeing the processes of designing, constructing, and operating nuclear reactors
- Overseeing the broader impacts of nuclear reactors such as community development, roads, utility, and telecommunication projects
- Start the process of building the necessary human infrastructure. In other words, training and education indigenous technocrats and workers in the field of nuclear power
- Initiating public information campaigns to educate and inform the public about the progress in the nuclear program.[28]

Additionally, the International Advisory Board (IAB) was established. Dr. Hans Blix, former director-general of the International Atomic Energy Agency, currently serves as the chair of the IAB and other members are top scholars and experts on nuclear power. The IAB's mission is to review the progress the nation is making in pursuing peaceful nuclear power and provide independent assessment of the performance of all institutions involved in the nuclear program.[29]

These three institutions – FANR, ENEC, and IAB – constitute the main nuclear power institutional infrastructure and are in charge of all nuclear power programs, operations, and policies. The UAE officials have been aware of the potential conversion of nuclear fuel to military uses and the political and security complications of enriching uranium. In order to avoid any suspicion or misunderstanding, they decided to import the nuclear fuel instead of claiming the right to make it. Furthermore, Abu Dhabi has been a strong supporter of an international, not national, approach to enriching uranium. In other words, instead of every country enriching the uranium it needs to empower nuclear reactors, the UAE has backed the proposal put forward by the Nuclear Threat Initiative, a non-governmental organization working on non-proliferation. This proposal calls on the international community to establish an international nuclear fuel bank from which countries can buy the fuel they need. The IAEA would manage the bank to ensure the non-interruption of supplies and that the process will not be politicized. In August 2008 Abu Dhabi pledged $10 million to support the establishment of the international nuclear fuel bank.[30]

In order to further alleviate any concern over nuclear weapons proliferation and to cement international support and trust, the UAE signed a number of international treaties and agreements. In addition of being a signatory of the Non-Proliferation Treaty (NPT), Abu Dhabi signed and joined the Additional Protocol to the Safeguards Agreement, the International Convention on the Suppression of Acts of Nuclear Terrorism, the Convention of the Physical Protection of Nuclear Material, the Convention on Nuclear Safety, the Joint Convention on the Safety of Spent Fuel Management and the Safety of Radioactive Waste Management, and World Association of Nuclear Operators.[31] The UAE also negotiated and signed a number of memoranda of understanding and bilateral nuclear energy cooperation agreements with the United Kingdom, Japan, South Korea, and the United States.

The UAE and the United States first signed a memorandum of understanding on nuclear energy cooperation in April 2008 then negotiated and signed a bilateral agreement in January 2009. Under this agreement

Abu Dhabi agreed not to 'enrich uranium or reprocess spent fuel'.[32] This is the first time that a country seeking to build nuclear power makes a legal commitment to forge enrichment and reprocessing.[33] The agreement permits 'the transfer of nuclear material, equipment, and components for civil nuclear research and civil nuclear power production'.[34] It also allows exchange of technical information and cooperation in education and training. The US government considers this agreement to be a model for other Middle Eastern countries (and elsewhere) to follow. Then US Secretary of State Condoleezza Rice, who signed the agreement on behalf of the US government, said the key to the deal is the UAE's 'willingness to import, rather than produce, fuel that would be used in its proposed reactors and that it would also return all spent nuclear fuel rather than attain the technical capability to reprocess it'.[35]

The US law (so-called Section 123) requires that any agreement should meet several non-proliferation criteria. These include safeguards on transferred nuclear technology and fuel, IAEA supervision, the US right to veto any transfer of nuclear material, physical security of nuclear installations, and ban on enrichment and reprocessing among others.[36] In order to ensure that all these non-proliferation criteria have been met, a US-UAE bilateral working group was established and meets annually to assess and address issues of mutual concern.

In addition to negotiating and signing memorandum of understanding and cooperation agreement, Abu Dhabi and Washington have worked together to start the process of launching the UAE nuclear program. The US firm CH2M Hill, Inc., won a contract to serve as the managing agent for the evaluation and design stage of the nuclear program and the Rizzo and Associates Inc. to survey potential nuclear reactor sites.[37]

In late 2000s several international companies competed to win a lucrative deal to build four nuclear reactors. The top three contenders were a French consortium led by Areva, an American-Japanese led by GE-Hitachi, and a South Korean led by Korea Electric Power Company (KEPCO), and involves Samsung, Hyundai, and Doosan, as well as Westinghouse. Eventually the latter was selected. Initially the construction costs were estimated at $20.4 billion but by late 2011 it was reported that they might reach $30 billion.[38] The UAE authority continued negotiations with the losing bidders, Areva and GE-Hitachi, regarding cooperation in related nuclear areas. The Korean consortium is expected to earn additional billions of dollars by jointly operating the reactors and by providing the nuclear fuel. In March 2010 KEPCO awarded a $5.59 billion construction contract to Hyundai and Samsung for the first plants.

The site chosen for the first nuclear complex is Braka, an uninhabited region far from the sea in order to avoid natural disaster such as tsunami or contaminating the water by accidently leaking radioactive materials. The groundbreaking ceremony was held in March 2011.[39] This 1,400 megawatt (MW) reactor is planned to become operational by 2017 and further add three units to generate a total of 5,600 MW by 2020.[40] The UAE is committed to a 'dual track' radioactive waste management strategy that involves developing a national storage and disposal program in parallel with exploring regional cooperation options. Abu Dhabi signed an agreement with the Swedish company SKB to study the prospects of a geological waste repository in the UAE.[41]

One of the UAE government's objectives is to develop well-qualified indigenous technocratic and managerial capacity to run the nuclear program. An important step in this direction was the establishment of the Gulf Nuclear Energy Infrastructure Institute (GNEII) in Abu Dhabi in 2010. The GNEII is a joint educational scheme between the UAE government and the Sandia National Laboratories and Texas A&M University in the United States. The GNEII's mission is 'to develop a responsible nuclear energy culture for the future decision-makers in the Gulf region, and institutionalize key safety, security, and non-proliferation norms in nuclear energy programs through professional development and training'.[42]

Renewable energy

Like nuclear power, renewable energy represents an insignificant proportion of the energy mix in almost all Arab countries. Recently most Middle Eastern countries have expressed interest in renewable energy. The reasons for this rising interest are similar to those of nuclear power, i.e. reduce dependence on oil, diversify the energy mix, meet surge in electricity consumption and demand, and protect the environment. As Sultan al-Jaber, chief executive of Masdar explained, 'The question is not to replace the oil and gas sectors, it's basically to extend the life of these two sectors and find alternative for the future'.[43] He added, 'We're hoping the UAE will become the Silicon Valley of renewable energy'.[44] Most of the planned schemes focus mainly, but not exclusively, on utilizing the region's solar power. This important source of energy is widely untapped and has the potential to make significant contribution to domestic consumption and electricity exports to neighbouring countries.

The UAE has taken the lead in utilizing solar power and other renewable sources. The goal is generate 6 to 7 per cent of the electricity from renewable energy by 2020.[45] By being at the forefront of renewable energy, UAE officials hope to position their country as a hub for solar power and major exporter of electricity and green technology. In pursuing such strategy, they have invested substantial resources in a variety of projects in recent years.

A key step was the creation of Abu Dhabi Future Energy Company in 2006, which is in charge of building Masdar City (source in Arabic). It is completely funded by the Mubadala Development Company, a UAE-owned investment firm. Masdar is designed to be a near-zero waste, zero carbon, and automobile-free city.[46] This huge project is estimated to 'cost $22 billion, serve as home to 50,000 people and 1,500 businesses and be powered mostly by solar energy'.[47] Masdar's website lists six objectives:

- Expand the export base
- Encourage private-sector entrepreneurship
- Invest in education and research that stimulates innovation
- Train, attract and retain skilled workers in knowledge-based sectors
- Encourage investment in areas that generate intellectual property gains
- Grow the non-oil sector's share of the Emirates' economy and decouple economic growth from fluctuating oil prices.[48]

Despite these high expectations, the creation of Masdar city has slowed down due to the global financial crisis, rise in expected costs, and frequent delays.

Shams 1 (Sun in Arabic) is one of Masdar's flagship projects. It is a joint venture between Masdar (60% of shares), Total SA of France (20%), and Abengoa Solar of Spain (20%), and is being developed on a 25-year build-own-operate (BOO) basis.[49] Upon completion and with a capacity of nearly 100 MW, Shams 1 will be the largest concentrated solar power (CSP) plant in the Middle East and one of the largest in the world.[50] When the sun's rays are unavailable, the plant will use natural gas to continue power production.[51] When Shams 1 becomes operational, Masdar plans to add Shams 2 and 3.[52]

Another important project is the Hydrogen Energy (HE), initiated in 2008. The HE is a joint venture between BP Alternative Energy and Rio Tinto, an Anglo-Australian firm. The proposed scheme will combine natural gas with carbon sequestration and hydrogen manufacture. Natural gas will be processed to create hydrogen and carbon dioxide

(CO_2). The hydrogen fuel would 'generate low-carbon electricity and the CO_2 would be captured and injected into mature oil fields to maintain pressure and keep them producing'.[53] Shortly after inception, this ambitious scheme has faced substantial uncertainties particularly regarding carbon credit and the UAE government's financial support. Consequently, Rio Tinto pulled out of the project and sold its 50 per cent shares to BP and the completion date moved from 2013 to 2014.[54]

Geothermal energy has been utilized in a few countries around the world, but rarely in the Middle East. In the late 2000s the UAE government has taken important steps to build a geothermal energy facility. It hired Reykjavik Geothermal, an Icelandic company as consultant. In 2010 Ensign, an Australian company, started drilling two wells.[55] The plan is for engineers at Masdar City to circulate water through the wells that should be 'heated into steam and used to turn electricity generators and directly power the City's air conditioning system'.[56]

As discussed above, the UAE seeks to educate and train indigenous technocrats and workers to manage and implement these ambitious renewable energy projects. In pursuing this goal, Abu Dhabi has been working closely with some of the top universities and research institutions in the world including the Imperial College in London, Tokyo Institute of Technology and Colombia University and the Massachusetts Institute of Technology in the United States.[57] One outcome of this collaboration was the founding of the Masdar Institute of Science and Technology (MIST). The institute's vision and mission are spelled out on its website:

> To be a world class graduate level institution seamlessly integrating research and education to produce future world leader and critical thinkers in advanced energy and sustainability and to position Abu Dhabi as a knowledge hub and engine for socio-economic growth.[58]
>
> Establish and continually evolve interdisciplinary, collaborative research and development capability in advanced energy and sustainability and educate students to be innovators with the breadth and depth to grow technology and enterprise in the region and globally.[59]

In recognition of the UAE's huge efforts and investment in pursuing renewable energy, the World Future Energy Summit has chosen Abu Dhabi as the location for its annual meeting in the last several years and the city was also chosen as the global headquarters of the International Renewable Energy Agency.

Conclusion

Three conclusions can be drawn from the discussion of the UAE energy outlook. First, like most of its neighbours, Abu Dhabi has legitimate reasons to pursue nuclear power and renewable sources. But more than the other Gulf Arab states, the UAE has taken several steps to implement these initiatives and to build the technological and human infrastructures. Thus, upon completion, the nuclear and renewable schemes have the potential not only to enhance the Emirates' energy security but to position it as a major regional hub and exporter of electricity and green technology to neighbouring countries.

Second, it is important not to get too optimistic and not to underestimate the enormous political, financial, and technological hurdles Abu Dhabi has to overcome in order to bring all these nuclear and renewable initiatives into fruition. The details of some of these projects are yet to be ironed out. For example, will a settlement of Iran's nuclear program (either by peaceful negotiations or military action) have any impact on the UAE's nuclear programme? All these programmes require sustainable strong financial support for many years to come. Will Abu Dhabi keep providing the necessary funding? The UAE holds substantial financial deposits and the Emirati leaders have expressed strong support to diversifying the energy mix. But, it is also true that the Emirates' economy is vulnerable to the fluctuation of oil prices. A decline in oil prices can weaken political and financial commitment. Despite close consultation and cooperation with the world's leading research institutions and universities, the UAE still lacks the necessary indigenous technocratic and managerial human power. Relying on foreign experts and labour has its own limitations and educating and training nationals will take a long time. In short, there is no guarantee that all these ambitious plans will be implemented. Indeed, as discussed above, progress in some of these projects has slowed down and the completion dates have been delayed.

Third, for decades, the UAE has been a major oil producer and consumer. In recent years the country has become a significant (per capita) consumer as well. In order to enhance the nation's energy security and maintain its status in the global energy markets both sides of the energy equation (production and consumption) need to be addressed. The focus in recent years has been mainly on the production side, i.e. adding other sources of energy, with little efforts made to curb consumption. One key reason for the high consumption level in the UAE and elsewhere is the very generous government subsidies. Subsidy exists when 'there is a financial contribution by a government or public body that confers a benefit'.[60] As a result, domestic petroleum prices are way below and do not reflect the market value. These artificially low prices encourage overconsumption and add to pollution. These subsidies need to be phased out and a culture of energy conservation and efficiency needs to be promoted.

7
Iran

In recent years very few issues have attracted international attention as the controversy over Iran's nuclear program. Iranian officials regularly remind both friends and foes of Article IV of the Treaty of the Non-Proliferation of Nuclear Weapons (NPT), which states, 'Nothing in this Treaty shall be interpreted as affecting the inalienable rights of all the parties to the Treaty to develop research, production and use of nuclear energy for peaceful purposes without discrimination'.[1] The United States and some other countries accuse Iran of seeking to build nuclear weapons. Diplomatic pressure and some of the severest economic sanctions have been imposed on Tehran combined with threats of military attacks. Iranian officials have categorically and consistently denied these accusations.

There is no way to speculate on how this conflict might end. However, this excessive focus on the Islamic Republic's nuclear program risks overlooking a comprehensive approach to the nation's energy consumption, production, and the overall outlook. The nuclear program is one component of the broad and complicated energy sector. Iran enjoys massive hydrocarbon deposits. It holds the world's third largest oil proven reserves (after Saudi Arabia and Venezuela) and second largest natural gas reserves (after Russia).[2] The question sceptics of Iran's nuclear program raise is: With all these hydrocarbon deposits, does Iran really need nuclear power? In response Iranian officials argue that their country's nuclear program is driven by strategic, economic, and environmental considerations.

- It is true that the nation holds considerable hydrocarbon resources. These deposits, however, are finite and cannot be replaced. High consumption and depletion rates further complicate the energy outlook.[3]

- All countries, consumers, and producers seek to improve their energy security by diversifying their energy mix. Oil and natural gas dominate Iran's energy sector. Developing and utilizing other sources of energy would enhance the country's energy security.
- Similarly, curbing fossil fuel consumption and increasing the share of nuclear (and renewable) power would reduce pollution. Tehran and other major Iranian cities are among the most polluted in the world.
- Financially, Iran is heavily dependent on oil revenues, which provide substantial proportion of the government's income and the overall national income. Diversifying the energy mix would free more oil to be exported instead of consumed domestically.[4] In addition, utilizing oil in petrochemical and other processing industries would secure remarkable added value for the economy.[5]

Rhetoric aside, analysts and policymakers can agree or disagree on these claims and counter-claims. Meanwhile, in the last few decades, Iran's energy sector has suffered from a combination of domestic and foreign pressure due to economic mismanagement, over-consumption, lack of international investment, war with Iraq, and unilateral and multilateral sanctions. Given the crucial role the energy sector plays in the broad economic system and, indeed, in securing domestic political stability, the Iranian government has taken several initiatives in recent years to vitalize its energy industry.

The next section highlights some of the main characters of the country's oil sector Together, these characteristics suggest that the authority needs to take decisive steps to curb consumption and increase production as well as reduce the heavy dependency on petroleum products. This will be followed by a close examination of the natural gas industry. Despite massive deposits, Iran is not a major gas exporter. Most of the production is consumed domestically. In recent years Tehran has invested huge financial and political capital in reducing dependency on fossil fuels and developing alternative energy. Renewable and nuclear power initiatives will be examined. Finally, any attempt to improve energy security has to address both the supply and demand sides. The concluding section analyses the efforts to enhance energy efficiency.

The oil sector

The Islamic Republic of Iran was the first country in the Persian Gulf where oil was discovered. Iran holds approximately 10 per cent of the world's proven oil reserves and is a founding member of the Organization of Petroleum Exporting Countries (OPEC). Tehran also enjoys a strategic location close to the lucrative energy markets of Asia and close to Europe with easy access to the high seas.

Despite these significant geological and geographical advantages, Iran's current status has remained well below its actual potential. Several political and economic factors have hindered Iran and slowed down the full utilization of its massive oil resources. These include the broad upheaval that accompanied the 1979 revolution and the subsequent Iran–Iraq War (1980–88), economic sanctions (imposed since 1979), a high population growth rate, and economic mismanagement. Furthermore, Iran faces continued depletion of its production capacity, as its fields have relatively high natural decline rates coupled with low recovery rate.[6] The combination of all these factors has negatively impacted economic performance, particularly the oil sector – the backbone of national economy.

The discovery and development of oil in Iran reflect three distinctive characteristics: Iran was the first country in the Persian Gulf and the Middle East where oil was found; the relations between the Iranian authorities and international oil companies in charge of petroleum exploration and development operations within the country were, generally speaking, characterized by mistrust and tension; and US involvement in the Iranian oil industry started in the mid-1950s, more than four decades after oil was discovered.

In 1901 the Shah of Iran granted a concession to William Knox D'Arcy, a British adventurer, to find, exploit, and export petroleum anywhere in Iran, except for the five northern provinces (Azerbaijan, Gilan, Mazanderan, Astrabad, and Khorassan), which were excluded as a result of Russian influence. Oil was first struck in 1908 in Masjid-i-Suleiman, on the site of an ancient fire temple.[7] The Anglo-Persian Oil Company was formed in 1909; it was renamed the Anglo-Iranian Oil Company in 1935, and the company is now known as British Petroleum (BP). The United Kingdom was particularly interested in discovering oil in Iran for at least three reasons: (1) there were no indigenous oil deposits in the United Kingdom nor in any part of the British Empire, as known at the time; (2) in 1913, shortly before World War I began, the Admiralty, then headed by Winston Churchill, decided to shift from coal to oil as fuel for the Royal Navy[8]; and (3) control of Iranian oil would strengthen a British presence and influence in the Middle East and deter the threat of German or Russian expansion in that area. Taking these reasons into consideration, the UK government bought a controlling interest in the Anglo-Persian Oil Company. Accordingly, Iranian oil facilities were rapidly expanded during World War I and by the early 1950s were still the best developed in the Persian Gulf region.

In spite of the continued expansion of the Iranian oil industry, tension and suspicion between the Anglo-Iranian Oil Company and the authorities in Tehran built up to a showdown in the early 1950s. The Iranian grievances focused on three areas: (1) the monopoly position enjoyed by the company; (2) the close relationship between the company and the UK government; (3) dissatisfaction with the financial terms of the concession between the company and the Iranian government. Developments between the early 1930s and the 1950s had underscored these Iranian complaints and contributed to the rise and sharpening of nationalism in Tehran. In 1932, the Iranian government decided to cancel the concession it had awarded to the Anglo-Persian Oil Company three decades earlier. After lengthy negotiations, the two sides signed a new concession in 1933 with more favourable provisions to Iran. Following this new agreement, the Iranians were satisfied with the steadily increasing income they were receiving and for a number of years had good relations with the company.

However, World War II introduced new dynamics. In 1941, the United Kingdom and the Soviet Union occupied Iran and forced Shah Reza, who was sympathetic to Germany, to abdicate in favour of his son. Several months later, Tehran, London, and Moscow signed a tripartite treaty of alliance. In 1948 and 1949, negotiations between the Iranian government and the Anglo-Iranian Oil Company produced an agreement supplementary to the 1933 concession, offering improved financial terms to Iran, though allowing little scope for the assertion of Iranian sovereignty over its oil resources.[9]

The Iranian Majlis (Parliament) rejected the agreement and passed the Nationalization Bill in April 1951, providing for the creation of a National Iranian Oil Company (NIOC), to which the assets of the Anglo-Iranian Oil Company were to pass and which would be the government agency responsible for running all aspects of the Iranian oil industry. This movement was led by Dr. Muhammad Mossadeq, who became prime minister. In response to these changes, Iranian oil production fell steeply and alternative oil resources were rapidly developed in the Persian Gulf, particularly in Kuwait, Saudi Arabia, and Iraq. The next two years witnessed an economic confrontation between London and Tehran, with Washington fearing that popular discontent and economic straits might lead Iran towards communism. These economic and political uncertainties came to an end with a coup d'etat, supported by the US government that led to the arrest of Mossadeq and the installation of a friendlier government in Tehran. Later, an agreement was signed between the new Iranian government and the Iranian oil participants, which was generally known as the Consortium.

The Consortium was initially made up of several international oil companies: British Petroleum, 40 per cent; Royal Dutch/Shell, 14 per cent; Gulf, Mobil, Standard of New jersey, Standard of California, and Texaco, each 7 per cent; Compagnie Francaise des Petroles, 6 per cent; and Iricon Agency, comprising eight US independent oil companies, 5 per cent.[10] The terms of the agreement were in line with the offers that had been made to, and refused by, Mossadeq at various times after the Act of Nationalization. Exploitation and marketing rights were assigned to the Consortium, but it was recognized that NIOC remained the sole owner of all fixed assets. Thus, in principle, the 1954 agreement recognized Iranian nationalization of the oil industry and provided what amounted to a 50/50 share of profits. In 1966 the Consortium relinquished 25 per cent of its agreement area, and in 1973 NIOC assumed control over oil production and refining in the whole agreement area.

To sum up, the process of expanding Iranian control over the ownership of its oil resources took place earlier and was more confrontational than the experience in other Persian Gulf producers. In spite of Iran's failure in the early 1950s to completely control its hydrocarbon wealth, by 1973 Tehran took charge. Still, NIOC allocated areas for exploration and development under joint arrangements with international oil companies and granted them service contracts.

Since its establishment during the Mossadeq crisis, NIOC has faced two often contradictory demands. On the one hand, oil is a fungible economic commodity that must be traded to be valuable. NIOC must sell in the international market to generate revenues for the Iranian national treasury. The state decides how to spend these funds based on both strategic and commercial interests. On the other hand, oil is not only economic commodity, but also seen as symbol of Iran's national strength and pride.[11] In this regard, NIOC serves as the national custodian of the country's most prized commodity. Due to Iran's confrontational encounters with international oil companies in the early decades of oil production, mistrust of foreign investment, and a strong resource-nationalism are more present than elsewhere in the region.[12]

The majority of Iran's crude oil reserves are located in the south-western Khuzestan region near the Iraqi border. One of the most recent and largest oil discoveries is the Azadegan field. It is located west of Ahvaz close to the Iraqi border and is considered one of the biggest discoveries in the world in the last few decades. The first exploration well was drilled in the field in 1976, but its discovery was finalized only after drilling a second well in 1999. Azadegan field is geologically complex and its oil is difficult to extract. Iran's reserves are not confined to the south-western region and offshore Persian Gulf. There are unconfirmed reserves offshore of the Caspian Sea. These reserves are largely underdeveloped due to territorial disputes with Azerbaijan and Turkmenistan as well as lack of funding.

Until recently Iran has limited refining capacity. The severe shortage of gasoline and other petroleum products have been recognized for several years. In February 1999 the Majlis passed an important law allowing foreign investors up to 45 per cent equity participation in Iranian oil refineries.[13] Also in the same year, the National Iranian Oil Refining, Production, and Distribution Company (NIORPDC) was set up by the oil ministry to supervise the nation's nine operating refineries. In order to address this severe shortage of refineries, the Iranian authorities have embarked on a large refining expansion and upgrading program as well as building new refineries. However, there has been little progress due to financing difficulties.

Natural gas

Iran's status as a natural gas producer and exporter and its place on the world gas stage do not accurately reflect its declared proven reserves.[14] In 2010 the country's share of the world's proven reserves was 15.8 per cent while its share of production was only 4.3 per cent.[15] This discrepancy can be explained by historical, geological, and strategic reasons. Traditionally, Iran has been an oil producer and exporter for more than a century. For a long time gas received a low priority because oil offered better and higher foreign exchange returns.

For most of the first half of the twentieth century associated gas was mainly flared. In early 1960s based on a contract signed with Russia, associated gas was gathered and transferred to the Soviet Union. These exports continued for several years until they were stopped due to price and payment disagreements.[16] Later, it was sent to Shiraz where a cement factory was the first factory to be fuelled by natural gas.[17] Gradually gas was supplied to the rest of the country and emerged as a major part of the energy mix.

Since the early 1990s, official attention has focused on developing gas resources. This was partly in response to the skyrocketing of domestic oil consumption, which cut into crude exports and threatened to deprive the state of a substantial proportion of annual revenues. The Iranian government responded by launching a gasification program in 1992. Under this program, hundreds of cities and towns were given access to gas.[18] Since then gas consumption and production have witnessed a steady increase. In the second decade of the twentieth century natural gas accounts for approximately 54 per cent of Iran's total domestic energy consumption.[19] This rising gas production is being utilized in at least four areas:

- Domestic use, including generation of electricity
- Gas injection into oilfields to enhance oil recovery
- Gas-based industries including petrochemical and gas-to-liquid (GTL) for domestic consumption and export
- Gas export to neighbouring countries.

A large share of Iran's gas production is consumed domestically. In 2001 Iran started exporting gas to Turkey and in 2009 to Armenia. Tehran also considered several proposals to export liquefied natural gas (LNG), but the lack of the necessary investment has delayed building terminals and other facilities. The Islamic Republic imports gas from Turkmenistan to meet the growing need of its northern region. Iran also has ambitious plans to export gas to its Arab neighbours on the Persian Gulf, to India and Pakistan, and to Europe. Very little success has been achieved due to lack of funding and international sanctions.

In order to manage the growing natural gas sector, the Iranian authority created the National Iranian Gas Company (NIGC) in 1965. The NIGC seeks to achieve several objectives including enhancing production, increasing export, encouraging private funding, attracting foreign investment, and expanding natural gas human and technical infrastructure.[20] The turning point in the nation's natural gas industry came in 1990 with the discovery of the South Pars.

Iran's South Pars field is the northern extension of Qatar's North Field. This structure is the biggest gas reserve in the world. It accounts for about 10 per cent of the world's confirmed reserves and 48 per cent of Iran's.[21] Originally the field was discovered by Shell in 1966, but major developments were delayed till early 1990s. South Pars is one Iran's biggest hydrocarbon projects. The field is being developed in approximately 25 phases about half of them had already been completed. In 1998 Pars Oil and Gas Company (POGC), a subsidiary of the NIOC, was established.[22] Its main mission is to manage the development of South Pars phases including oil, gas, petrochemical industries, and refining facilities.[23]

The development of South Pars has been slow partly because of international sanctions and partly due to the unattractive financial and legal terms Iran has offered in the last two decades. Following the 1979 revolution, a new constitution was established, stipulating that foreign ownership of the country's natural resources was illegal. Specifically, it prohibits the granting of petroleum and gas rights on a concessionary basis or direct equity stake. The oil and gas industry was nationalized, the government expropriated the assets of foreign companies operating in the country, and concession agreements were broken. Eight years later, when the attitude towards foreign investment started to change, the Petroleum Law of 1987 was promulgated. It permits the establishment of contracts between the Ministry of Petroleum and state companies on one side and local and foreign natural citizens and legal entities on the other.[24] Since reopening to the international oil companies (IOC), Iran has employed the buy-back framework for upstream contracts. The adoption of this principle represents a radical break with the past and was a courageous move that carried huge political risks for those who espoused it.

The buy-back model is better explained in comparison to other formulas that have been utilized between oil producers and foreign companies. These include the joint venture (JV), the production-sharing agreement (PSA), and the technical services contract (TSC). JVs are arrangements under which a state enterprise and a foreign firm invest stated amounts of capital that can take various forms, including funds, intellectual property, or physical assets and rights to land. The partners share the risk and the reward of the venture in proportion to the capital contributed. Under PSAs, the foreign contractor is reimbursed for exploration, development, and operating costs by way of a certain share of production.[25] In other words, the foreign partner holds no equity stake but does maintain a firm legal entitlement to a certain percentage of oil and gas volume. Finally, TSCs involve a simple cash payment for services rendered. These types of contracts entail little risk, if any, for the foreign firms. Subsequently, the rewards do not rise should the discovery be substantial or oil prices increase.[26] In addition, TSCs tend to be awarded for short periods.

The buy-back formula does not violate the Iranian constitution. It is a service contract under which one or more parties are contracted by the Ministry of Petroleum to carry out necessary exploration and development work on a field that, once completed, reverts wholly to the Ministry. Thus, the foreign company is neither a partner nor a concessionaire, but acting in the role of a hired contractor servicing the national company.

The buy-back model demands that the foreign partner provide all the investment capital for exploration and, in return, is paid a predetermined rate or return on capital invested. This is paid in kind after the production of the first commercial oil. Buy-back contracts generally are designed by the Iranian negotiators to last five to seven years and are thus fairly short term in the context of the traditional upstream contract.[27]

Although there have been several agreements signed between Iran and IOCs based on buy-back formulas, the model has been criticized particularly for three reasons. First, it is almost risk-free for the IOCs, since they are guaranteed a return on their investment. Second, the brief duration of the contracts does not provide incentives for the IOCs to introduce and fully utilize their advanced technology and management skills. Third, the fixed rate of return leaves IOCs with little incentives to maximize the profit from oil and gas fields. In short, the criticism focuses on the disconnection between performance and profit and the absence of long-term relationships between performance and profit and the absence of long-term relationships between the Iranian energy authority and IOCs.[28]

With all these shortcomings taken into consideration, there have been intense debates in Tehran and negotiations with IOCs regarding modification of the buy-back model. New contracts signed with IOCs link performance to payments and provide for long-term relations with foreign companies. The goal is to optimize oil and gas field utilization. A consensus is growing in Iran that, without modifying the buy-back formula, the country could fall short of its expectations in terms of attracting the necessary foreign investment. This is particularly important to offset how some foreign investors see Iran as a political risk due to US, EU, and UN sanctions.

Iran's rising gas domestic consumption is projected to further grow to meet its large population. Also the share of natural gas used for reinjection in old and mature oil fields is expected to increase dramatically. Meanwhile, economic sanctions and the absence of major international companies have dealt a blow to the natural gas industry and significantly slowed down production. Against this background, Tehran has shown growing interest and determination to develop renewable and nuclear energy.

Renewable energy

Iran's energy mix is overwhelmingly dominated by fossil fuels. Indeed, the entire economy (residential, industrial, and transportation sectors) runs almost exclusively on oil and natural gas. Hydroelectricity represents a small, but growing, proportion of the nation's energy sector. This heavy dependency, as the discussion in the previous section underscores, has proven unsustainable. The country has considerable human and natural resources to modernize its energy supply and manage a transition to a more sustainable energy system. There are favourable conditions for broad utilization of the so-called green energy.[29] In the last few decades the Iranian authorities have sought to develop the nation's untapped renewable energy resources, particularly geothermal, wind, and solar.

Geothermal energy: Iran has promising geothermal energy potential. The country is located on the world volcano-seismic belt and consequently has experienced several organic phases during geological times, which left a lot of semi and non-active volcanoes.[30] According to one study about 8.8 per cent of Iran is considered prospected geothermal areas.[31] The list of the main regions for geothermal energy generation includes Sabalan, Makoo, Khoy (Azerbaijan Province), and Damavand (Tehran Province).[32] The roots of geothermal projects go back to the mid-1970s when James R. McNitt, a United Nations geothermal expert visited Iran. The following year the Ministry of Energy, Entes Nazionale per L'Energia Elettrica of Italy, and Tehran Berkeley of Iran signed an agreement for geothermal exploration in the north-western part of Iran.[33] After years of extensive geological studies, the parties identified four locations (Sabalan, Damavand, Khoy-Maku, and Sahand) as the most proper as geothermal sites. More potential areas were added following a country-wide geothermal energy resource exploration survey carried out by the Renewable Energy Organization of Iran (SUNA), Kyushu University, and Sinclair Knight Merz Ltd of New Zealand. Researchers at the Kyushu University also developed digital geothermal potential map of Iran using geographic information system (GIS).

Following extensive surface exploration and feasibility studies, a few wells were drilled at Sabalan. The goal was to explore and evaluate subsurface geological conditions, geothermal reservoir assessment, and response simulation. The results were positive. As a result, the government installed several geothermal heat pumps in different parts of the country and invested resources to educate the public about this source of energy. These pumps serve both as air conditioning and water heating systems. They produce very little pollution.

The most prominent geothermal project is the Sabalan Power Plant located on the north-western portion of Sabalan volcano, near Meshkinshahr city in the province of Ardebil, north-west Iran.[34] Iranian engineers drilled several wells. The goal is to convert heat from the earth into electricity. The plant is the first geothermal power development in both Iran and the broader Middle East.[35]

To sum up, Iran holds several potentially high-grade geothermal prospects. In the last four decades it has invested in building technological and technical infrastructure. This infrastructure has been built as a partnership between Iranian scientists and engineers and their counterparts in foreign countries. The full and speedy utilization of the country's geothermal potential requires the continuation of this partnership with foreign universities and research institutions. It also requires steady private and foreign investment. International sanctions have negatively impacted the country's progress in developing this clean source of energy. Finally, geothermal energy faces stiff competition not only from fossil fuels, but also from other renewable resources.

Wind: Despite boasting some of the best wind resources in the Middle East, Iran's wind energy market has not yet realized its full potential. Some of the earliest known windmills in the world were in ancient Iran. In recent times considerable efforts have been made to develop wind energy. In early 2000s several stations were set up to record data on wind potentials and to determine the precise wind capacity in the country. Based on this date the country's first wind atlas was compiled. Most of the wind corridors are located in the mountainous part of the country, along the Alborz and Zagros mountain chain. At the beginning of the second decade of the twenty-first century, Iran's wind power sites are concentrated in four sites: Manjil wind farm, Binalood wind farm, Loutek wind site (Tabriz), and Eouibnali wind site (Zabol).[36] Additionally, in mid-2011 the country's first wind-solar hybrid power plant in Kharg Island in the south was inaugurated.[37] Wind energy is being used largely to generate electricity and pump water, among other uses.

Despite the slow growth of wind energy industry, the infrastructure capacity for rapid deployment is well-established. The authority seeks not only to diversify the nation's energy mix, but also to develop the indigenous industry into a wind energy hub for the entire Middle East and West Asia. Indeed, Iran exports wind technology to neighbouring Armenia and Pakistan and has negotiated deals with other neighbours.

Solar: Iran's solar electricity generation is tremendous. This can be explained at least by two geographic facts. First, Iran is one of the largest countries in the Middle East. Its desert occupies roughly one quarter of the total land area. Second, the country receives almost eight hours a day of sunshine or 3,000 a year, one of the highest in the world. This means that the average solar radiation is about 19.23 Mega joules per square meter, again one of the highest in the world.[38] Some of the best sites for solar thermal power plants are Esfahan, Fars, Kerman, and Yazd.

In recent years a number of solar power plants went on line. In the late 2000s there are thousands of small-scale solar thermal installations and residential solar water heaters throughout the country.[39] In January 2009 the Shiraz solar power plant was inaugurated. This solar thermal plant uses parabolic mirrored troughs to gather sunlight. The mirrors focus the sunlight in an intense ray on a tube that runs the length of the array of mirrors. Inside the tube, a liquid insulated by a vacuum transfers the heat of the mirrors to a traditional generator, where it is used to produce steam and generate electricity. The plant adds 250 kW of solar energy to the country's grid.[40] It was constructed with domestic materials and labour in Shiraz (Fars province).

In July 2011 another solar power plant was inaugurated in the north-eastern city of Mashhad. It makes use of 216 solar panels and is equipped with solar trackers that deliver solar energy production, generating 72,000 kW-h of electrical power each year.[41]

As a sign of strong official support to renewable energy the Iranian government established the Renewable Energy Organization of Iran (SUNA) in 2000. The organization seeks to contribute to the stabilization and diversification of energy resources, the development of human resources, reduce long-term expenses associated with energy generation, and preservation of environment.[42] The research on renewable resources is carried out by several universities and research institutions including Iranian Research Organization for Science & Technology, Materials and Energy Research Center, University of Shiraz, University of Sharif, University of Tehran, and University of Meshad, among others.

To conclude, Iran has tremendous potential to diversify its energy mix, reduce the heavy dependency on fossil fuels, and slow and contain pollution. These objectives can be realized if geothermal, wind, and solar powers are utilized. In the last few decades Tehran has made some progress towards achieving these goals. However, the confrontation with the United States and other major powers over the country's nuclear program has reduced foreign investment and slowed the development of renewable energy.

Nuclear power

For the last several years Iran's nuclear program has been one of the most controversial issues in the Middle East and the world. Iran's pursuit of nuclear power is seen as an example, a threat, or a combination of both by most states in the region.[43] Since the early 2000s Turkey and several Arab states have declared their intention to build nuclear reactors and a few have already taken first steps including signing agreements with foreign partners and allocating substantial financial resources. With the exception of Iran, which is developing enrichment and reprocessing capability and Israel, which has reprocessing capability, no other state in the Middle East has expressed an intention to develop such fuel cycle activities.[44]

Given the dual-use character of both nuclear technology and nuclear materials, Iran's nuclear program has been increasingly viewed more as a security/strategic issue and less as a technical energy one. On one side, some major global powers, led by the United States claim that with its massive oil and natural gas resources Iran does not need nuclear power. On the other side, the Iranian government (seemingly supported by many Iranians) argues that other major energy producers such as Canada and Russia have utilized nuclear power and that the Islamic Republic, as a signatory of the NPT, has similar rights. The Iranians further claim that Western powers try to deny them such rights.

The initiation and evolution of Iran's nuclear program reflect the ups and downs of Tehran relations with global powers, particularly the United States. Like many other countries, Tehran sought foreign technology and expertise. The Iranians also have always been interested in developing indigenous technological capabilities. Thus, the foundations of the nuclear program were laid down in partnership with American, French, and German governments and research institutions.

Initially, the United States was hesitant to sell nuclear technology to other countries out of fear that this technology might be used to make weapons. The Atomic Energy Act (or the MacMahon Act) passed by the US Congress in 1946 highlighted this concern. Other Western powers (mainly Britain and Canada) took advantage of US policy of restricting the transfer of technology. Only American companies were not allowed to sell nuclear technology, while British and Canadian companies, among others, were very active.

Against this background, Washington changed its policy. President Dwight Eisenhower's speech 'Atoms for Peace' before the United Nations General Assembly in December 1953 signalled this drastic departure. Under certain conditions and guarantees, American companies became active players in the global market seeking to help allies of the United States to build peaceful nuclear power. In addition to these commercial interests, then Washington was a relatively new player in the Middle East. The United States played a key role in overthrowing the nationalist

Iranian Prime Minister Mohammad Mossadeq in 1953. This was a significant beginning of a decades-long troubled relationship. In the following 16 years, the Shah was considered among the strongest allies the United States had in the region, and probably the world.

After short negotiations, Washington and Tehran signed an agreement on civilian nuclear cooperation in 1957 (in 1969 it was extended for another 10 years). The agreement gave birth to Iran's nuclear industry and called for the U.S. Atomic Energy Commission to lease Iran low-enriched uranium (LEU) for research purposes. Two years later, the Shah ordered the establishment of Tehran Nuclear Research Center (TNRC) at Tehran University and negotiated with the United States to supply a five-megawatt thermal research reactor for the centre.[45] Cooperation between the two sides ensued and Iran received the reactor as well as other laboratories.

The Shah was keen to build and develop a large pool of Iranian nuclear engineers and scientists. In the 1960s and 1970s hundreds of Iranian students were sent to study nuclear physics and engineering in top universities around the world, particularly in the United States and Europe. In addition, several nuclear research centres and departments were created. Given the Shah's strong ambition, the authorities generously funded these scholarships and centres.

In the late 1960s/early 1970s two developments provided the Shah with the necessary legal cover and financial resources to substantially push forward the nuclear program. First, Tehran was one of the first countries in the world to sign and ratify the NPT. It also completed its Safeguards Agreement with the International Atomic Energy Agency (IAEA). As a signatory, the Treaty gives Iran the right to develop peaceful nuclear power. Second, the 1973 Arab-Israeli war, followed by the first oil shock (when Arab oil producers reduced production and imposed oil embargo on the United States and a few European powers), led to a surge in oil prices. Iran accumulated billions of dollars from oil revenues. This massive wealth provided the necessary funding to the nation's then infant nuclear program. In addition, Iran's demand and consumption of electricity was surging, reflecting the country's growing economic development.

All these developments led the Shah to accelerate the building of the country's nuclear program. In 1974 the Shah established the Atomic Energy Organization of Iran (AEOI) and announced plans to more than double Iran's nuclear-generated electricity in the following two decades. The AEOI has since been considered the main official body responsible for implementing regulations and operating nuclear energy installations. It has facilities throughout the country and is divided into different divisions, including nuclear power plant, nuclear regulatory authority, research division, and nuclear fuel production division.[46]

In the following years several agreements were signed with the United States, Germany, and France, among others to build nuclear power plants (NPP). Washington agreed to build eight NPPs, to exchange nuclear technology, and to cooperate on nuclear safety. Meanwhile the French firm Framatome agreed to build two pressurized water reactors at Darkhovin near Ahvaz. Finally, Germany agreed to build six nuclear power reactors, and the first two were to be built by the firm Kraftwerk Union (a subsidiary of Siemens) at Bushehr. The Shah also increased the AEOI budget by several folds.[47] Furthermore, the Iranian authority signed a contract with the Massachusetts Institute of Technology (MIT) to train Iranian engineers. Similar contracts were signed with French and German research institutions. The training of Iranian nuclear engineers in the West, signalled the Shah's determination to build a strong technical infrastructure. Another sign was the founding of the Nuclear Technology Center at Esfahan. In short, given the Shah's close alliance with the West, the United States and European states did not hesitate to provide Iran with nuclear technology and fuels.

The dramatic events of the late 1970s and early 1980s, including the toppling of the Shah regime and the return of Ayatollah Khomeini, the beginning of the war with Iraq, and the hostage crisis, dealt a heavy blow to the nation's infant nuclear program. Then, Iran had two plants under construction by Germany's Siemens at Bushehr.[48] Several developments contributed to this huge setback. First, the strong anti-Shah and anti-American sentiments in the early days of the revolution led to the dismantling of many projects associated with the old regime, including the nuclear program. Second, Ayatollah Khomeini was concerned that the partnership with US and European firms would make Iran too dependent on foreign technology. Third, the political and economic turmoil forced many Iranian nuclear scientists and engineers to leave the country. Iran lost a large number of its best young nuclear professionals in what is known as 'brain-drain'. Fourth, the combination of domestic instability and the war with Iraq led to massive cuts in the country's oil revenues. The Islamic Republic had less financial resources to fund the nuclear program. Fifth, the deteriorating relations with the United States and economic and diplomatic sanctions did not only stop cooperation between the two sides, but also Washington exerted tremendous pressure on other countries to freeze their nuclear programs with Iran.

All these developments led major nuclear projects to be cancelled or to be left dormant.

In the late 1980s the Iranian authority had to confront a changing landscape both domestically and internationally. Despite the war and sanctions, the nation had one of the fastest population growth rates. This swelling population needed more electricity. At the same time, Iraq attacked the Bushehr site several times and destroyed most of the two reactors. In addition, the eight-year war and the economic sanctions took their tolls on the country's hydrocarbon production. Oil infrastructure (both upstream and downstream) suffered from lack of the necessary investments. Finally, Washington put heavy pressure on its allies (and other countries) to stop any nuclear cooperation with Tehran. European governments and firms were unwilling to support Iran's nuclear program.

This combination of rising demand and shrinking supply left Iran with a few options. The need to sell more crude and to diversify the energy mix made the nuclear power more attractive. The efforts to convince several European firms from Argentina, Germany, Spain, Italy, among others, to complete the Bushehr reactors failed. Similar attempts to enlist Chinese cooperation did not succeed. In 1990, then President Akbar Hashemi Rafsanjani reached an agreement with his Soviet counterpart Mikhail Gorbachev under which the Soviet Union agreed to complete the reactors.[49] However, the collapse of the Soviet Union shelved this agreement. A few years later (1995), seeking to boost its civilian nuclear industry and export, Russia agreed to finish (or more accurately to re-build from scratch) the Bushehr reactors. After several delays, the reactor became operational in 2012 and was connected to the electrical grid.

This brief discussion of Iran's efforts to build and develop nuclear power suggests two conclusions. First, Iran (both under the Pahlavi and Islamic regimes) has not been satisfied with just importing nuclear technology and reactors. The Iranians have been determined to build indigenous technical capacity. Since the late 1950s Tehran has sent its young scientists and engineers to study nuclear technology wherever possible. Equally important, several research centres and new departments were created inside Iran to teach and train nuclear technology. Based on available data from open resources, it seems Iran has created a more advanced technological and human capacity than most of its neighbours. Second, with few exceptions, the Iranians have been adamant on reducing their vulnerability to foreign supplies and insisting on enriching uranium and/or reprocessing plutonium.

This insistence on full fuel cycle started under the Shah and has continued till today. Iranians have always perceived their nation as a regional power (and even a global power). As such Iran should enjoy the same status and rights like other advanced countries that produce nuclear fuel. The list includes not only nuclear weapons states, but others such as Argentina, Brazil, Germany, Japan, the Netherlands, and South Africa.[50] Nuclear fuel can be exported to other countries and add another source of public revenues and improve trade deficit. Finally, Iran's history and psychic indicate a great deal of scepticism and lack of trust in foreign powers.

Iran's experience with reliance on foreign supplies of nuclear fuel further reinforces this skepticism. In 1973 Belgium, France, Italy, Spain, and Sweden formed a joint stock company called EURODIF (European Gaseous Diffusion Uranium Enrichment Consortium). The idea behind this consortium was to create an enrichment facility in France and deliver the enriched uranium to all the partners. Instead of each country enriching its own needs, a large-scale production would serve all partners' interests. Since France is a nuclear-weapons state, diversion of enriched uranium to support a clandestine nuclear weapons program was not an issue.

Shortly after the consortium was established, Sweden withdrew and its 10 per cent share went to Iran as a result of an arrangement between the French and Iranian governments. In 1974 the Shah lent $1 billion (and another $180 million in 1977) to the French Atomic Energy Commission for the construction of a plant to enrich uranium. The loan would have entitled Iran to buy 10 per cent of enriched uranium produced by EURODIF. Shortly after the 1979 revolution, Iran cancelled its agreement with EURODIF and halted payments for enriched uranium deliveries. After a bitter legal dispute, Iran was reimbursed its original loan plus interest. Though Iran remains an indirect shareholder of the consortium and demanded delivery of enriched uranium based on the old contract, France has refused to deliver the enriched uranium. Iranian leaders argue that this experience demonstrates the unreliability of foreign suppliers and the necessity of having a full fuel cycle.[51]

In August 2002 a representative of the National Council of Resistance of Iran held a press conference in Washington, DC, and revealed that Iran has a uranium enrichment facility in Natanz and a heavy water production facility in Arak.[52] This news was alarming to Tehran's rivals for at least two reasons. First, enrichment uranium and heavy water may be used for either peaceful or military purposes. Second, the news indicated that Iran has made significant progress in developing its nuclear program. Iranian officials have repeatedly denied that they have a secret plan to build nuclear weapons or that they violated the NPT. They, however, acknowledged that 'they failed to report some of the progress in their nuclear program to the IAEA in a timely fashion'.[53]

Since then the controversy over Iran's nuclear program has intensified with Western power, led by the United States accusing Iran of seeking to make nuclear weapons while Tehran categorically denies these accusations and asserts that it is only interested in peaceful nuclear power. The two sides have held several rounds of negotiations and the United Nations Security Council issued four resolutions (1696, 1737, 1747, 1803) imposing sanctions on Iran. Meanwhile, Iran has continued making progress in developing a full fuel cycle.

Given the dual-use character of nuclear technology and material, Iran's nuclear program has become a major regional and international controversy. Interestingly both Iran and some of its adversaries exaggerate nuclear achievements. The former would like the world to believe that its mastery of the fuel cycle is already a fait accompli.[54] The latter seeks to build a case for a military action. Based on open sources, it is hard to make accurate assessment of where Iran's nuclear program stands. Still some closing remarks can be made.

First, nuclear power would improve Iran's energy security, diversify its energy mix, divert millions of barrels of oil from domestic consumption to export, and reduce pollution and contribute to the efforts to contain global warming and climate change. An Iranian analyst argues that the current plans to produce 20,000 MW of nuclear electricity by 2020 'may save Iran 190 million barrels of crude oil every year, or nearly $14 billion annually'.[55] Second, some analysts claim that the scale of facilities required to reach market competitiveness in nuclear technologies is 'inconsistent with the scale of Iran's uranium resources'.[56] According to this line of analysis, given the relative small size of Iran's nuclear program, multilateral approach to secure nuclear fuel might be more cost-effective than a national one.

Finally, regardless of economic calculus, enriching uranium and indeed the whole nuclear program has taken on too much political weight. The very strong domestic support and national pride cannot be neglected. In addition, since the late 1950s Iran has amassed significant technical knowledge and engineering skills. This experience and knowledge cannot be destroyed by economic sanctions or military strikes. They are transferable.[57] All parties need to reach an agreement to ensure that nuclear technology and materials will not be used to make weapons.

Conclusion: Iran energy outlook

Any long-term comprehensive planning for energy outlook should address the two sides of the energy equation – supply and demand. Iran is blessed with massive hydrocarbon deposits and long experience in oil and gas industries. However, Iran's energy sector faces serious challenges. These include mismanagement, high rate of natural depletion, lack of foreign investment, lack of access to the most advanced technology, and economic sanctions among others. As a result, more than three decades after the revolution, oil production has not reached the peak of the mid-1970s. Similarly, despite having the second largest natural gas proven reserves in the world, Iran is not a major gas exporter. Iran needs to increase its oil and gas production and further expand its investment in alternative energy (nuclear and renewable).

On the demand side, like its Arab oil-producing neighbours, Iran has one of the highest levels of energy intensity in the world. In 1996 the Iranian authority created Iran Energy Efficiency Organization to promote consumption rationalization in all sectors.[58] Four years later (2000), Iranian Fuel Conservation Company, a subsidiary of National Iranian Oil Company, was established. Its mission is to 'regiment the fuel consumption in different sectors through review and survey of the current trend of consumption and executing conservation measures nationwide'.[59]

Energy waste and smuggling are encouraged by a system of price subsidies for gasoline, natural gas, and electric power. After long political debate in December 2010, the Iranian government started a comprehensive system of energy reform by reducing subsidies and allowing prices to move closer to market values. It is too early to assess the success of this policy, but it seems a step in the right direction and a positive development to curb consumption.

8 Conclusion

Energy plays a crucial role in modern societies. It has a vital input to all sectors (e.g. residential, transportation, and manufacture) and is essential to generate electricity. In other words, all societies require energy services to meet basic human needs such as lighting, heating, and mobility. Thus, energy is not just a regular commodity, rather it is a strategic one and is the lifeblood of today's civilization.

In the last several decades the world was heavily dependent on coal, oil, and natural gas to meet its energy needs. To be sure, these fossil fuels still dominate the energy mix in most countries today and are projected to continue this domination in the foreseeable future. However, the oil embargo and the interruption of supplies that followed the 1973 Arab–Israeli war (so called first oil shock) sent a wake-up call not only to Western countries, but to the entire world. The clear message was oil supplies were vulnerable to geopolitical changes. The repeated threats or interruptions of gas supplies from Russia to Ukraine and other countries in the last two decades further underscored this vulnerability. In addition, recent scientific data and growing global consensus suggest that consumption of fossil fuels accounts for the majority of global anthropogenic greenhouse gas (GHG) emissions. Stated differently, fossil fuels, particularly coal, are major polluters.

These geopolitical vulnerabilities and environmental risks were the main drivers behind the global search for alternative energy. Indeed, the use of alternative energy is inevitable as fossil fuels are finite. Nuclear power occupies a unique position in the debate over global climate change as the only carbon-free energy source that is already contributing to world energy supplies on a large scale and that is also expandable with few inherent limits.[1] The impact of the March 2011 nuclear accident in Fukushima, Japan, is yet to be seen. In the aftermath of this tragedy some countries suspended or scaled back their nuclear programs and others decided to keep their strong nuclear ambition. Similarly, renewable energy can provide numerous environmental, economic, and social benefits. In addition to the reduction of carbon dioxide emissions, governments have initiated policies to utilize renewable energy to meet several objectives. These include advancement of energy security goals by diversifying the portfolio of energy technologies and resources; facilitation of electricity access, particularly for rural areas; and improving social and economic development through potential employment opportunities.

In the Middle East some countries like Iran, Saudi Arabia, and the United Arab Emirates (UAE) hold substantial proven hydrocarbon reserves. Upon development, the recent natural gas discoveries in the Eastern Mediterranean are likely to drastically improve Israel's energy outlook and security. In recent years Egypt's oil production has declined but the country has emerged as an important gas producer and exporter. Egypt's large and fast-growing population, however, means more gas will be consumed domestically and less will be available for export. Finally, Morocco holds very limited indigenous hydrocarbon reserves and imports most of its needs.

Despite this variation, these six Middle Eastern states share environmental concerns and ambitions to utilize nuclear technology and renewable resources. Generally, the region has a large potential for renewable resources, particularly solar power. These resources, however, are largely untapped. Some of the case studies examined in this volume have the necessary financial resources to initiate nuclear programs. The demand for both electricity and primary fuels in the six countries is rapidly rising. According to a recent study by the World Bank, since 1980 the consumption of energy has grown faster in the Middle East/North Africa than in any other region in the world.[2] One reason for this high growth rate of energy consumption is the low level of energy efficiency. Another reason is the artificially low prices of energy supported by heavy public subsidies.

The initiation and development of alternative energy in the Middle East and improving energy efficiency require comprehensive strategy to encourage private investment, to build the necessary institutions and enact the appropriate laws and regulations, and to partner with foreign governments and corporations. All the six countries examined in this volume have, in different degrees, addressed these issues. As a result, the energy landscape in the Middle East is fundamentally changing. Essentially, there is a realization that the heavy dependency on oil and natural gas has to be contained, making room for nuclear and renewable energy. Meanwhile, every available option to meet the demand for energy has its own limitations. On the demand side, all the six countries have addressed energy saving and efficiency. Certainly, there is a long way to go to reach compatible levels to those of other countries.

After examining the similarities and differences between the six case studies, the analysis in this chapter will focus on how to slow the rise in energy consumption, mainly by reducing subsidies and improving efficiency. Particular attention will be given to the Iranian initiative since 2010. Finally, both nuclear and renewable energies require regional cooperation. The Seven Countries Interconnection Project (SCIP) and the Gulf Cooperation Council (GCC) grid will be examined as examples of joint efforts by neighbouring countries to utilize their resources.

Energy security

For long time the world relied on fossil fuels to meet most of its energy needs. Not only were the prices affordable, but equally important interruption of supplies triggered by political disputes was not an issue. The creation of the Organization of Petroleum Exporting Countries (OPEC) in 1960 as a cartel representing the interest of major oil-producing countries served as a sign that a key change was about to take place. The so-called oil embargo (1973–74) that followed the 1973 Arab –Israeli war represented a turning point in the decades-long perception of energy security. Since then the supplies of oil and natural gas have been interrupted due to political crises. These geo-strategic disputes have also contributed to intense price fluctuation and volatility.

Against this background the interest in alternative energy has emerged and evolved since the mid-1970s. Problems related to safety, reliability, and affordability have slowed down the maturation of nuclear and renewable power. In recent years technological advances and lower costs have convinced many countries to take a fresh look at these alternative energy resources. Furthermore, growing environmental concerns have made energy security inseparable from the transition to a low-carbon economy.[3] These developments have broadened the perception and understanding of energy security. In 2011 the International Energy Agency (IEA) adopted a comprehensive approach that includes availability (geological), accessibility (geopolitical), affordability (economic), and acceptability (environmental and social).[4] Finally, the availability of reliable supplies at affordable prices with little environmental impact represents only half of the energy equation. The other half is efficient demand. Stated differently, energy security has supply-side and demand-side components.[5]

Within this context, the six countries examined in this volume have pursued similar strategies to enhance their energy security. First, the initiation of alternative energy programs was driven by similar reasons (e.g. the desire to diversify the energy mix, reduce pollution, free more oil to be exported instead of being consumed domestically, and the need to meet the surge in energy consumption, particularly electricity). In addition, Saudi Arabia, the UAE, and Morocco lack adequate water supplies and depend on desalination to meet their water needs. This process is energy intensive. Second, energy is a top priority in the six countries. One ministry or more is in charge of the energy sector. For example, the Moroccan Ministry of Energy, Mines, Water and Environment and the Saudi Ministry of Petroleum and Mineral Resources supervise and regulate the energy sector in their respective countries. In Israel, these responsibilities are divided between several ministries led by the Ministry of National Infrastructure. In all six countries the private sector and foreign investors have recently taken a bigger and growing role in the energy sector in partnership with the state. De-regulating the energy sector and creating an attractive environment to attract foreign investment are guiding principles in the six countries.

Third, the availability of a competent technical infrastructure to support research and development varies from one country to another. For instance, Israel has numerous energy research institutions, most of them are affiliated with universities and are actively involved in joint research projects with their European and American counterparts. Iran and Egypt also have several research institutions focusing on the energy sector. In recent years Saudi Arabia and the UAE have invested massive financial resources to educate and train indigenous technocrats and carry out research in different disciplines including energy. King Abdullah City for Atomic and Renewable Energy and King Abdullah University of Science and Technology are good examples. King Fahd University of Petroleum and Minerals (established in 1963) has been leading energy training and research in the kingdom for several decades.

Fourth, in recent years all the six countries have expressed interest in nuclear power. Iran has the most advanced program with the most recent nuclear plant (Bushehr) going critical in early 2012. Egypt has had two nuclear plants for decades and Israel has a research centre at Nahal Soreq since the 1960s. Since the late 2000s the UAE has taken significant steps to build nuclear reactors and Saudi Arabia also seems determined not to fall behind. Meanwhile, Morocco has expressed interest but taken little action to initiate a nuclear program. Similar to other countries, nuclear programs in the six countries depend heavily on assistance from foreign companies (Russian, South Korean, French and others). In order to assure the international community that these nuclear programs are solely for peaceful applications, the six countries have signed, ratified, and joined several treaties and agreements on nuclear safety and non-proliferation.

Fifth, almost the entire Middle East enjoys tremendous renewable energy resources potential. These include abundance of sunny days in combination with huge arid zones. In addition, some of the world's best wind resources are available in Iran, Egypt, and Morocco. Geothermal energy potentials are limited, available mainly in Iran. Hydropower makes good contribution in power generation in Iran, Egypt, and Morocco. The combined share of renewable resources in the energy mix, however, is fairly limited.

Sixth, the Middle East region has one of the highest energy intensity and one of the highest consumption per capita in the world. The authorities in the six countries examined in this volume have been increasingly aware of the risks associated with this high consumption and in recent years have launched several initiatives to curb consumption and promote conservation. In 1996 the Iranian government founded the Energy Efficiency Organization, in 2009 Egypt created Energy Efficiency Unit, and a year later Saudi Arabia established Center for Energy Efficiency and Israel introduced National Energy Efficiency Program. Meanwhile, the Moroccan government requires all its agencies to design an energy efficiency strategy. Finally, in all six countries efforts have been made to engage civil society and promote public awareness.

To sum up, fundamental changes are likely to re-shape the energy landscape in the Middle East in the foreseeable future. Most of these changes have been initiated since the early 2000s or earlier. The main characteristics of this evolving energy landscape include: First, nuclear power still is in infant stage, most countries lack the necessary technical and human infrastructure, while some (e.g. the UAE and Saudi Arabia) have the financial resources to launch and fund ambitious nuclear programs, others (e.g., Egypt and Morocco) have harder time allocating these resources. Finally, important issues such as fuel cycle and waste management have yet to be addressed. Second, renewable resources are widely available in the region. However, they are largely under-exploited and under-utilized. In recent years all the six countries have taken initiatives to diversify their energy mix. The full utilization of renewable resources, however, will take some time and will depend on the building of adequate institutions, enacting appropriate laws and regulations, engaging private sector and encouraging foreign investment. Finally, fossil fuels, particularly oil and petroleum products and to a less extent natural gas, still dominate the energy mix in all six countries. Their share is likely to decline slowly as nuclear and renewable resources make gains. Subsidies, efficiency, and regional cooperation are likely to impact the evolution of energy strategies in the six countries.

Subsidies

The high level of energy consumption in the Middle East and North Africa cannot be divorced from the region's strong tradition of underpricing energy. In most countries of the region, subsidies for fuels and electricity constitute a significant share of government spending. Indeed, Arab countries, particularly oil producers, are among the largest energy subsidizers in the world. To be sure, most countries in the world subsidize energy in one way or the other. According to the IEA, fossil fuel consumption subsidies worldwide amounted to $409 billion in 2010, up from $300 billion in 2009, with subsidies to oil products representing almost half of the total.[6] Recent analyses by the Organization for Economic Cooperation and Development (OECD) and the IEA indicate that phasing-out fossil fuel subsidies could lead to a 10 per cent reduction in global GHG emissions in 2050 compared with business-as-usual.[7]

Policymakers and energy analysts have adopted different stances on the controversy over subsidies. Yousef Alyousef and Paul Stevens define a subsidy as the 'difference between the market price and the real opportunity cost of the commodity'.[8] The Joint Economic Committee of the US Congress describes it as 'any government assistance, in cash or in kind, to private sector producers or consumers for which the government receives no equivalent compensation in return, but conditions the assistance on a particular performance by the recipient'.[9]

The World Trade Agreement on Subsidies and Countervailing Measures provides a detailed and comprehensive definition. According to Article 1 a subsidy exists when there is a financial contribution by a government or public body that confers a benefit. A financial contribution arises where (a) a government practice involves a direct transfer of funds (e.g. grants, loans, and equity infusion); (b) government revenue that is otherwise due is foregone or not collected (e.g. fiscal incentives such as tax credits); (c) a government provides goods or services other than general infrastructure, or purchases goods; or (d) a government entrusts or directs a private body to carry out one or more of the above functions.[10]

The argument for energy subsidies in the Middle East (and elsewhere) is based on several environmental, economic, political, and strategic grounds. (a) Initiating, developing and marketing renewable and nuclear power usually require government financial support at least in the early stages. All the six case studies examined in this volume depend heavily on fossil fuels and are among the most polluted countries in the world. Providing public financial incentives is likely to protect their infant alternative energy programs, mitigate environmental problems, and diversify their energy mixes; (b) Subsidies proved crucial to supply some rural areas in Iran, Saudi Arabia, Egypt, and Morocco with electricity and primary fuels. This development has contributed to economic development, poverty reduction, and job generation; (c) Some of the Middle Eastern countries are critically vulnerable to the fluctuation of oil prices in the global market and its impact on prices in the local economy. Subsidies help to maintain the prices of electricity, natural gas, and petroleum products low and protect against inflation; (D) In major oil-producing countries, particularly Saudi Arabia and the UAE, energy subsidies are part of a broad and un-written social contract between the authority and the people. Under such contract, the authority ensures the supply of energy and other basic commodities at low prices and in return people demand less in terms of political participation and transparency.

Opponents of subsidies argue that artificial low fuel prices create distortive price signals. These low prices are indiscriminate, all energy consumers, rich and poor, benefit from the availability of cheap electricity and petroleum products.[11] But having bigger homes and cars, the wealthy individuals and firms consume more and benefit more. Furthermore, cheap fuel prices encourage over-consumption and waste. They also create disparity between domestic market and neighbouring ones. For a long time heavily subsidized petroleum products were smuggled to neighbouring countries and sold at higher prices. Higher consumption also generates more airborne emissions and greenhouse gases.[12] Finally low prices of oil and gas products leave few incentives to invest and develop alternative energy resources.

All energy sources receive some forms of state support. Indeed, hydropower, wind, and solar (and other renewable sources) would be nowhere near viable without some forms of subsidies. However, worldwide and particularly in the Middle East renewable energy receives less money in annual subsidies than fossil fuels do.[13] These heavily subsidized fossil fuels make it harder for alternative energy to compete in terms of costs.

Despite these relative disadvantages, a recent report by the IEA concluded that while renewable electricity remains generally more expensive than conventional sources, renewable deployment is 'starting to transition from a phase in which it is more reliant on subsidy support to one in which projects are competing on their own merits'.[14]

This debate over subsidies has intensified in recent years and in response, the authorities in several Middle Eastern countries have considered or started implementing strategies to address resource misallocation and reduce subsidies. As the case studies illustrate, new laws and regulations have been enacted in Morocco, Egypt, and particularly Iran, to bring petroleum products in line with market prices. In Israel after long debate, new tax regulations were issued governing the natural gas sector.

The efforts to reform fuel prices have faced new complications since the early 2011 due to political upheavals in the Arab world, the so-called Arab Spring. In Saudi Arabia and the UAE the authorities responded to popular discontent by increasing public spending. The rise of populist leaders in Egypt suggests that reducing fuel subsidies (which will certainly lead to higher prices) is not a priority of the new government in Cairo. In short political uncertainties in much of the Arab world is likely to slow down any efforts to reform energy prices. The experience in Iran, however, may provide some lessons to its neighbours.

In 2005 President Mahmoud Ahmadinejad was elected and promised to implement a populist agenda, meaning helping the poor and implying the continuation of heavy subsidies. However, Iran's domestic gasoline price was 'clearly out of touch with reality, unsustainable and unjustifiable by any economic theory',[15] according to a recent study by the International Monetary Fund (IMF). Iran, a major oil producer, was importing increasing amounts of gasoline to supply domestic demand and make-up for large volume of crude and petroleum products being smuggled to neighbouring countries.

In 2007 Iran implemented a gasoline rationing system, which allowed consumers to buy a certain volume of fuel at the old subsidized price and anything above that at a much higher price. This new policy was not adequately explained to the people and the response was widespread public protest. Three years later, the authority introduced a more gradual and comprehensive reform. Officially, known as Targeted Subsidies Reform, the program made Iran the first major energy producing and exporting country in the Middle East to cut drastically heavy subsidies to energy products and replace them with cash transfers to the population. The scheme contains an element of re-distribution in favour of low-income households. The goal of these cash transfers was to discourage some marginal gasoline consumption and allow consumers to buy more other goods and services.

In June 2011 IMF officials congratulated Iran on the early success in implementing the subsidy reform program, describing it as 'a critical step in enhancing energy efficiency, supporting growth and moving further towards a market-based economy'. They also mentioned that cash transfers were 'instrumental in supporting domestic demand, improving income distribution and reducing poverty'.[16]

Efficiency

As the analysis of the six case studies demonstrated, investing and developing alternative energy sources (e.g. nuclear and renewable) are likely to enhance energy security, reduce pollution, and contribute to economic development and poverty reduction. However, the two sides of the energy equation – supply and demand – need to be addressed simultaneously. As the IEA's executive director, Ms. Maria Van der Hoeven argues, 'the best way to ensure energy security is to use less energy'.[17] In addition to improving energy security, efficiency is a good business that offers high returns on investments, increases the sustainability of energy sources and reduces pollution.

Most Middle East states, particularly major oil and gas producers, are among the world's least energy-efficient consumers. Energy intensity is measured as energy input per unit of gross domestic product (GDP) and energy consumption per capita. In the last few decades income growth in the Middle East has lagged other regions. Energy consumption, in contrast, has soared. In other words, energy consumption has been rising faster than GDP. Between 1990 and 2010 global energy intensity decreased by 1.2 per cent per year and fell in all regions except the Middle East.[18] In 1970, energy use per capita was roughly twice the rest of the non-OECD; by 2010, it was more than three times as high. In 1970 energy intensity was less than half the level of other non-OECD; by 2010, it was 50 per cent higher. With the trend in other countries pointing towards continuous improvement, British Petroleum (BP) predicts that the Middle East region is likely to be more than twice as energy intensive as the rest of the non-OECD by 2030.[19] In 2009 a study by the World Bank concluded that the Middle East's energy intensity was 60 per cent higher than that of the OECD countries and 40 per cent above the world's average.[20]

Several factors have contributed to this high level of energy intensity. As discussed above, low energy prices lead to overconsumption and waste. Another reason is the heavy concentration on energy-intensive industry (such as petrochemicals and aluminum) in the Gulf states. In recent years Saudi Arabia has sought to establish itself as the world leader in petrochemical industry. Low energy prices and energy-intensive industry reinforce each other and lead to inefficiency and push intensity higher. Two other technical factors contribute to the high level of energy intensity: gas flaring in oil and gas production, and transmission and distribution losses in the power system. The Middle East region ranks very high in both areas. In short, the high and growing energy intensity is due to the fact that the region's development is based on energy-intensive industries, as well as energy-intensive lifestyles in buildings and transport, encouraged by low energy prices.

These high levels of energy consumption and low levels of efficiency are unsustainable. They eat into the region's hydrocarbon reserves and export capability. The authorities in all the six case studies have realized these risks and started articulating and implementing strategies to address them. Thus, energy efficiency institutions and agencies have been established and laws and regulations have been enacted. Some of the common themes in these strategies need to be highlighted. First, technologies can play an integral role in transforming the energy system, reduce intensity, and enhance efficiency. Existing technologies make it possible to reduce dependency on fossil fuels (increase the volume available for export in Iran, Saudi Arabia, and the UAE and reduce imports in Morocco, Egypt, and Israel). Available technologies can also contribute to de-carbonizing electricity and reducing emissions in the manufacture, transport, and buildings sectors.[21]

Second, the establishment of several energy research institutions particularly in the UAE, Saudi Arabia, Iran, and Israel underscore the official attention given to the role of research and development (R&D) efforts in energy saving and diversification. What is missing, however, is a regional joint strategy to avoid duplication, improve performance, reduce costs, and share lessons.[22] Third, partnership with foreign governments and institutions has proven crucial in launching and consolidating nuclear and renewable energy programs. This means that it is important to create the adequate political framework to encourage investment from foreign and private corporations. Fourth, institutions and regulations are important. However, the success of a comprehensive energy strategy requires an active participation of the civil society. A campaign to educate and engage the public, development of education curricula, promotion of professional and industrial associations, and partnership with scientific and academic institutions are all important components of a successful energy efficiency strategy.

Regional power grids

The IEA projects that the annual average growth in global power generation from renewable sources accelerates to 5.8 per cent over the 2011–17 period versus 5.0 per cent in 2005–11.[23] The Middle East is likely to follow this trend particularly with the expansion of regional power grids such as the SCIP and the GCC Power Grid.

Cross-border power grid interconnections provide links between the electricity transmission systems of two or more adjoining countries, and accordingly, interconnections allow those countries to share power generation resources. The primary reason for developing an electricity grid interconnection between countries is to reduce the overall combined economic costs of supplying electricity services in the interconnected countries. Sharing resources can allow the construction of larger facilities with lower unit costs. In addition, grid interconnections can allow generating units with lower environmental impacts to be used more often while units with higher impacts are used less.[24] They also can provide diversity to the energy supply. For instance, instead of relying exclusively on oil, natural gas, and coal, the interconnections can increase system efficiency by importing electricity from nuclear or hydro grids to the receiving countries. In the case of certain fuels or resources such as hydropower and other renewable resources, interconnections are the only feasible means of making such resources available to other areas leading to the development of these diverse energy resources for benefits of the entire region, thus allowing less costly power to be delivered from distant locations, often displacing important, expensive fossil fuels, and utility projects. In short, they can contribute to environmental protection.

One of the most ambitious cross-border schemes is the SCIP. Launched in the early 1990s, SCIP aims to interconnect the grids of Libya, Egypt, Jordan, Syria, Iraq, Turkey, and Lebanon. The future plan is to link the SCIP with Europe via Turkey and Morocco. The SCIP should reduce power generation, improve efficiency, and enhance the overall energy security. Egypt's grids were connected to those of Libya and Jordan in 1998.[25] Each partner of the interconnected system must satisfy certain conditions in its own supply zone. For example, each company must help to compensate for a disturbance in the grid in proportion to its share of the total power output of the interconnected network at the time of the disturbance.[26]

Another ambitious interconnection project is the GCC Power Grid. Since its founding in 1981, the six members in the GCC – Bahrain, Kuwait, Oman, Qatar, Saudi Arabia and the United Arab Emirates – have sought to establish and consolidate social, economic, and political cooperation. After extensive negotiations and numerous feasibility studies, they created the Gulf Cooperation Council Interconnection Authority (GCCIA) in 2001 with the mandate to implement an interconnected grid. The GCCIA, a joint Gulf stock company, is located in Dammam, Saudi Arabia. The GCC grid is being implemented in three phases:

1. Interconnection of the Northern Systems (Bahrain, Kuwait, Qatar, and Saudi Arabia)
2. The internal interconnection of the Southern Systems (Oman and the UAE)
3. Interconnection of the Northern and Southern Systems.

The GCC grid aims at providing a platform for energy trade and exchange, while improving the reliability of existing energy systems and lowering electricity reserve requirements. Stated differently, the participation of private and foreign investments contributes to lower production costs and ultimately lower electricity prices. Additionally, these private and foreign investors are involved in the development of larger projects with access to larger markets.[27] In short, by providing sustainable energy supplies, the GCC grid will contribute to broad economic prosperity across the Gulf states. Another purpose of the interconnection is to share generation reserves and installed capacity in order to reduce additional investments in generation infrastructure.[28] There are talks about expanding the power grid beyond the GCC and possibly connect it to the rest of the Middle East and North Africa and even to Europe.[29] Such connections will provide an opportunity for the export of surplus power from the GCC in the mild winter (when demand is low) to Europe and from the latter to the former in the hot summer (when demand is high).

Conclusion: the way ahead

In recent years Middle Eastern countries, like the rest of the world, have realized that the continuation of heavy dependency on fossil fuels is unsustainable for a variety of reasons. In addition of being finite, the high level of consumption means less oil and gas are available for export. Furthermore, low-carbon energy system necessitates more diverse energy sources and entails increased efficiency. This change from a conventional energy system (where oil and gas dominate) to an alternative one (where the energy mix is more diversified) is driven largely by technological advances. The integrated use of key existing technologies has already made it possible to reduce dependency on fossil fuels, lower the costs of nuclear and renewable power, and enhance energy efficiency. In addition, the slow transition from fossil fuels to alternative ones requires aggressive policies to stimulate changes in the energy production and consumption.[30]

In the post-Fukushima era, rising costs associated with enhanced safety measures must be overcome. Similarly, in the Arab Spring era governments need to increase transparency in their decision-making processes and implement updated safety and risk management protocols. Equally important, independent nuclear regulatory bodies are required for industry oversight. Meanwhile, recent technological advances have made renewable resources more cost-competitive. Indeed renewable resources have been increasingly seen as a valuable component of any secure and sustainable energy economy, providing energy at a low cost with high price stability and minimum environmental footprint. Alternative energy is projected to play a growing role in meeting rising demand, particularly in power generation.

Despite these advantages (and challenges) of nuclear and renewable energy, fossil fuels will continue to dominate the energy mix in the Middle East in the foreseeable future. Oil and gas production is projected to rise with gas increasingly and persistently replacing oil particularly in power generation. This projected and deliberate rise in gas consumption is meant to leave more oil for export in countries like Iran, Saudi Arabia, and the UAE and to leave enough oil to meet rising consumption and declining production in Egypt and to replace imported fuels in Israel.

In recent years all the six countries examined in this volume and other Middle Eastern states have not only expressed interest in developing alternative energy, but also have launched several ambitious initiatives and programs to diversify their energy mixes. To be sure, every energy source has its own limitations and nuclear and renewable powers are likely to supplement, not supplant, fossil fuels.

Appendix

Alternative Energy: The Middle East and the World

Country	All Renewables	Wind Index	Onshore Wind	Offshore Wind	Solar Index	Solar PV	Solar CSP	Biomass	Geo-thermal
China	70.6	77	79	70	64	67	47	60	51
United States	69.4	68	71	57	74	74	75	63	69
Germany	66.4	69	67	80	61	70	0	69	58
India	65.5	66	71	43	67	69	54	62	45
Italy	58.4	59	61	50	60	62	41	53	61
UK	55.8	64	61	80	42	48	0	58	36
France	55.6	58	59	54	52	56	29	57	33
Canada	53.8	63	67	46	40	46	0	50	36
Brazil	49.8	53	56	40	45	46	33	54	24
Australia	49.5	49	51	38	53	53	54	43	57
Spain	49.2	48	51	36	54	53	60	44	27
Sweden	48.9	55	55	53	37	42	0	58	35
Japan	48.8	46	48	39	57	61	27	39	46
Romania	47.6	53	56	39	40	45	0	45	41
South Korea	46.6	47	46	53	48	51	29	41	36
Poland	46.2	53	56	41	36	42	0	43	22
Greece	44.9	45	48	33	49	51	33	34	25
South Africa	44.8	49	52	36	42	41	50	36	34
Portugal	44.4	46	48	34	46	47	35	38	25
Belgium	44.3	51	49	58	36	42	0	39	27
Mexico	43.0	44	45	39	43	43	40	39	54
Netherlands	42.9	49	49	49	37	42	0	37	21
Denmark	42.4	46	44	58	35	40	0	46	33
Ireland	42.1	52	52	50	26	30	0	43	23
Taiwan	41.3	44	45	39	39	44	0	37	38
Morocco	41.3	39	42	26	48	47	54	38	21
Norway	40.0	48	48	46	26	29	0	45	30
Ukraine	39.8	39	41	27	40	46	0	46	32
Egypt	39.8	42	44	31	39	39	44	34	24
Turkey	39.8	41	43	32	38	40	28	35	41
Austria	39.7	35	39	0	44	50	0	50	33
Finland	39.6	46	48	39	24	28	0	53	26
New Zealand	39.5	47	50	37	27	31	0	34	51
Tunisia	36.8	36	38	27	44	44	48	20	27
Bulgaria	36.5	36	39	24	37	42	0	35	34
Israel	36.1	33	37	14	45	46	38	26	28
Argentina	34.9	37	40	22	33	36	17	32	27
Hungary	34.0	34	38	0	31	35	0	43	39
Chile	33.4	35	38	23	32	34	19	28	36
Czech	32.6	34	38	0	31	36	0	31	23

Source: Ernst & Young, Renewable Energy Country Attractiveness Indices. Available at http://www.ey.com/publication/vwluassets/renewable_energy_attractiveness_indices_issue_33/$file/ey_recal_issue_33.pdf. (Accessed May 30 2012).

World Nuclear Power Reactors: December 2012

	Nuclear electricity generation		Reactors operable		Reactors under construction		Reactors planned	
Country	**kWh**	**%**	**No.**	**MWe**	**No.**	**MWe**	**No.**	**MWe**
Argentina	5.9	5.0	2	935	1	745	1	33
Armenia	2.4	33.2	1	376	0	0	1	1060
Bangladesh	0	0	0	0	0	0	2	2000
Belarus	0	0	0	0	0	0	2	2400
Belgium	45.9	54.0	7	5943	0	0	0	0
Brazil	14.8	3.2	2	1901	1 1	405	0	0
Bulgaria	15.3	32.6	2	1	906	0	0	1 950
Canada	88.3	15.3	20	14169	0	0	2	1500
Chile	0	0	0	0	0	0	0	0
China	82.6	1.8	15	11881	29	30010	52	60880
Czech Republic	26.7	33.0	6	3764	0	0	2	2400
Egypt	0	0	0	0	0	0	1	1000
Finland	22.3	31.6	4	2741	1	1700	0	0
France	423.5	77.7	58	63130	1	1720	1	1720
Germany	102.3	17.8	9	12003	0	0	0	0
Hungary	14.7	43.2	4	1880	0	0	0	0
India	28.9	3.7	20	4385	7	5300	18	15100
Indonesia	0	0	0	0	0	0	2	2000
Iran	0	0	1	915	0	0	2	2000
Israel	0	0	0	0	0	0	0	0
Italy	0	0	0	0	0	0	0	0
Japan	156.2	18.1	50	44396	3	3036	10	13772
Jordan	0	0	0	0	0	0	1	1000
Kazakhstan	0	0	0	0	0	0	2	600
North Korea	0	0	0	0	0	0	0	0
South Korea	147.8	34.6	23	20787	4	5205	5	7000
Lithuania	0	0	0	0	0	0	1	1350
Malaysia	0	0	0	0	0	0	0	0
Mexico	9.3	3.6	2	1600	0	0	0	0
Netherlands	3.9	3.6	1	485	0	0	0	0
Pakistan	3.8	3.8	3	725	2	680	0	0
Poland	0	0	0	0	0	0	6	6000
Romania	10.8	19.0	2	1310	0	0	2	1310
Russia	162.0	17.6	33	24164	10	9160	24	24180
Saudi Arabia	0	0	0	0	0	0	0	0
Slovakia	14.3	54.0	4	1816	2 880	0	0	
Slovenia	5.9	41.7	1	696	0	0	0	0
South Africa	12.9	5.2	2	1800	0	0	0	0
Spain	55.1	19.5	8	7448	0	0	0	0
Sweden	58.1	39.6	10	9399	0	0	0	0
Switzerland	25.7	40.8	5	3252	0	0	0	0
Thailand	0	0	0	0	0	0	0	0
Turkey	0	0	0	0	0	0	4	4800
Ukraine	84.9	47.2	15	13168	0	0	2	1900
UAE	0	0	0	0	1	1400	3	4200
UK	62.7	17.8	16	10038	0	0	4	6680
USA	790.4	19.2	104	102195	1	1218	13	15660
Vietnam	0	0	0	0	0	0	4	4000
World	2518	13.5	436	374135	65	65159	168	185495

Sources: World Nuclear Association, World Nuclear Power Reactors. Available at http://www.world-nuclear.org/info/reactors.html. Accessed 26 December 2012.
Operable = Connected to the grid.
Under Construction = first concrete for reactor poured or major refurbishment under way.
Planned = Approvals, funding or major commitment in place, mostly expected in operation within 8–10 years.

Notes

1 Introduction

1. ExxonMobil (2012) The Outlook for Energy: A View to 2040. Available at http://www.exxonmobil.com/energyoutlook. Accessed 4 August 2012.
2. British Petroleum (2012) *BP Statistical Review of World Energy* (London) pp. 6, 22.
3. Project on Managing the Atom and Russian Research Center Kurchatov Institute (2010). Promoting Safe, Secure, and Peaceful Growth of Nuclear Energy. Available at http://belfercenter.ksg.harvard.edu/files/promoting-safe-secure-and-peaceful-growth-of-nuclear-energy.pdf. Accessed 18 November 2010.
4. International Atomic Energy Agency (2010). International Status and Prospects of Nuclear Power. Available at http://www.iaea.org/about/policy/GC/GC54/GC54infDocuments/English/gc54inf-5_en.pdf. Accessed 2 September 2010.
5. WMD Insights (2010) Gulf Cooperation Council Moves Forward with Nuclear Energy Plans. Available at http://www.wmdinsights.com. Accessed 27 December 2010.
6. World Nuclear Association (2011) Three Mile Island Accident. Available at http://www.world-nuclear.org/info/inf36.html. Accessed 14 January 2011.
7. World Nuclear Association (2011) Chernobyl Accident. Available at http://www.world-nuclear.org/info/chernobyl/inf07.html. Accessed 14 January 2011.
8. Mycle Schneider, Antony Froggatt, and Steve Thomas (2011), *2010–2011 World Nuclear Industry Status Report* (Washington, DC: World Watch Institute), p. 61.
9. Institute of Nuclear Power Operations (2010) About us. Available at http://www.inpo.info/aboutus.htm. Accessed 12 December 2010.
10. World Association of Nuclear Operators (2010) About us. Available at http://www.wano.info/about-us. Accessed 27 December 2010.
11. International Atomic Energy Agency (2010) IAEA/NEA Incident Reporting System Using Operational Experience to Improve Safety. Available at http://www.iaea.org/publications/Booklets/IaeaNea/iaeanea-irs.html. Accessed 27 December 2010.
12. United Nations Security Council (2010) United Nations Security Council Resolution 1540 (2004). Available at http://www.un.org/sc/1540. Accessed 27 December 2010.
13. Nuclear Threat Initiative (2010) The Global Initiative to Combat Nuclear Terrorism: Progress to Date. Available at http://www.nti.org/e_research/e3_global_initiatives.html. Accessed 27 December 2010.
14. Nuclear Threat Initiative (2010) NTI in Action: World Institute for Nuclear Security (WINS). Available at http://www.nti.org/b_aboutnti/b7_wins.html. Accessed 27 December 2010.
15. Steven E. Miller and Scott D. Sagan (2010) 'Alternative Nuclear Futures,' *Daedalus* (139), 126–137.
16. Matthew Bunn (2009) 'Reducing the Greatest Risks of Nuclear Theft and Terrorism', *Daedalus* (138), 112–123.

17. European Commission (2010) Commission Staff Working Document Summary of the Impact Assessment Accompanying Document to the Revised Proposal for a Council Directive (Euratom) on the Management of Spent Fuel and Radioactive Waste. Available at http://ec.europa.eu/energy/nuclear/waste_management/doc/2010_11_03_proposal_directive_radiactive_waste_ia.pdf. Accessed 19 November 2010.
18. World Nuclear Association (2010) Processing of Used Nuclear Fuel. Available at http://www.world-nuclear.org/info/inf69.html. Accessed 21 October 2010.
19. Robert H. Socolow and Alexander Glaser (2009) 'Balancing Risks: Nuclear Energy and Climate Change,' *Daedalus* (138), 31–44.
20. World Nuclear Association (2010) Processing of Used Nuclear Fuel.
21. Mary Beth Mikitin, Anthony Andrews, and Mark Holt (2010) *Managing the Nuclear Fuel Cycle: Policy Implications of Expanding Global Access to Nuclear Power* (Washington DC: Congressional Research Service), p.29
22. International Atomic Energy Agency (2010) International Status and Prospects of Nuclear Power.
23. Yury Yudin (2010), *Multilateralization of the Nuclear Fuel Cycle: The Need to Build Trust* (Geneva: United Nations Institute for Disarmament Research), p. 43.
24. More information is available at http://www.igdtp.eu
25. Tom LaTourrette, Thomas Light, Debra Knopman, and James T. Bartis (2010) *Managing Spent Nuclear Fuel: Strategy Alternatives and Policy Implications* (Santa Monica, CA: RAND), p. xii.
26. Allison Macfarlane (2011), 'It's 2050: Do You Know Where Your Nuclear Waste Is?' *Bulletin of the Atomic Scientists* (67), 30–36.
27. International Atomic Energy Agency (2010) Joint Convention on the Safety of Spent Fuel Management and on the Safety of Radioactive Waste Management. Available at http://www.iaea.org/Publications/Documents/Conventions/jointconv.html. Accessed 27 December 2010.
28. International Energy Agency (2010), *World Energy Outlook* (Paris), p. 8.
29. Paul L. Joskow and John E. Parsons (2009) 'The Economic Future of Nuclear Power,' *Daedalus* (138), 45–59.
30. International Atomic Energy Agency (2010) International Status and Prospects of Nuclear Power.
31. Richard K. Lester and Robert Rosner (2009) 'The Growth of Nuclear Power: Drivers and Constraints,' *Daedalus* (138), 19–30.
32. Harold A. Feiveson (2009), 'A Skeptic's View of Nuclear Energy,' *Daedalus* (138), 60–70.
33. Olli Heinonen (2011), ' A Multinational Fuel Consortium: Obstacles, Options, and Ways Forward', *Bulletin of the Atomic Scientists* (67), 37–42.
34. Anatoly S. Diyakov (2010), 'The Nuclear Renaissance and Preventing the Spread of Enrichment and Reprocessing Technologies: A Russian View', *Daedalus* (139), 117–125.
35. Yury Yudin (2010), Multilateralization of the Nuclear Fuel Cycle: The Need to Build Trust, p. 34.
36. Director General IAEA (2010) Annex 20/20 Vision for the Future. Available at http://www.iaea.org/publications/magazines/bulletin/bull501/50101010608.pdf. Accessed 20 October 2010.
37. The text of the Non-Proliferation Treaty is available at http://www.un.org/Depts/dda/WMD/Treaty. Accessed 24 December 2010.

38. Steven E. Miller and Scott D. Dagan (2009), 'Nuclear Power Without Nuclear Proliferation?' *Daedalus* (138) 7–18.
39. Mohamed I. Shaker (2010), 'Nuclear Power in the Arab World and the Regionalization of the Nuclear Fuel Cycle: An Egyptian Perspective', *Daedalus* (139), 93–104.
40. The full text is available at http://www.fissilematerials.org/ipfm/site_down /ach46.pdf. Accessed 24 December 2010.
41. Thomas Lorenz and Joanna Kidd (2010) 'Turkey and Multilateral Nuclear Approaches in the Middle East', *Nonproliferation Review* (17), 513–530.
42. Jessica Jewell, 'A nuclear-powered North Africa: Just a Desert Mirage or Is There Something on the Horizon?' *Energy Policy* (39), 4445–4457.
43. Ottmar Edenhofer, Ramon Pichs Madruga, and Youba Sokona (2012) *Renewable Energy Sources and Climate Change Mitigation* (Cambridge: Cambridge University Press), p. 178.
44. International Energy Agency (2010) *World Energy Outlook* (Paris), p. 9.
45. British Petroleum (2012) BP Energy Outlook 2030. Available at http://www.bp.com. Accessed 30 January 2012.
46. Lloyd's Sustainable Energy Security (2010) Strategic Risks and Opportunities for Business. Available at http://www.chathamhouse.org.uk. Accessed 8 May 2010.
47. International Energy Agency (2012), Tracking clean energy progress: energy technology perspectives 2012 excerpt as IEA input to the clean energy ministerial. Available at http://www.iea.org. Accessed 28 April 2012.
48. Noriko Fujiwara, Monica Alessi, and Anton Georgiev (2012), Carbon market opportunities in the Middle East and North Africa, Mediterranean prospects. Available at http://www.medpro-foresight.eu. Accessed 24 March 2012.
49. Robert Mabro (2006) 'Renewable Energy', in *Oil in the 21st Century: Issues, Challenges and Opportunities* (London: Oxford University Press), pp. 327–343 (p. 337).
50. International Energy Agency (2011) *Deploying Renewables 2011: Best and Future Policy Practice*, (Paris), p. 15.
51. International Energy Agency (2010) Renewable Energy Essentials: Hydropower. Available at http://www.iea.org. Accessed 16 November 2010.
52. International Energy Agency (2012) *Medium-Term Renewable Energy Market Report* (Paris), p. 12.
53. For more information see International Hydropower Association (2011) About IHA. Available at http://www.hydropower.org/about_iha/introduction.html. Accessed 17 January 2011.
54. Ottmar Edenhofer, Ramon Pichs Madruga, and Youba Sokona (2012) *Renewable Energy Sources and Climate Change Mitigation* (Cambridge: Cambridge University Press), p. 71.
55. Josef Auer (2010) Geothermal energy: construction industry a beneficiary of climate change and energy scarcity, Deutsche Bank Research. Available at http://www.dbresearch.com. Accessed 23 February 2010.
56. International Energy Agency (2012), *Tracking Clean Energy Progress* (Paris), p. 27.
57. Robert Mabro (2006) *Renewable Energy*, p. 336.
58. Suzanne Hunt (2008) 'Bio-fuels, Neither Savior nor Scam: The Case for a Selective Strategy', *World Policy Journal* (25), 9–17.

59. Ralph Sims and Michael Taylor (2008), From 1st to 2nd Generation Bio-Fuel Technologies, International Energy Agency. Available at http://www.iea.org. Accessed 20 November 2008.
60. Lloyd's, Sustainable Energy Security (2010) Strategic Risks and Opportunities for Business.
61. Global Wind Energy Council (2011) Regions. Available at http://www.gwec.net. Accessed 17 January 2011.
62. International Energy Agency (2012) Tracking Clean Energy Progress. Available at http://www.iea.org. Accessed 28 April 2012.
63. Energy Information Administration (2010) How Much of the World's Electricity Supply Is Generated from Wind and Who Are the Leading Generators? Available at http://www.eia.doe.gov. Accessed 23 June 2010.
64. Anne Feltus (2008), 'Solar Power's Day in the Sun', *Petroleum Economist* (51), 20–22.
65. Global Energy Network Institute (2007), Renewable Energy Potential of the Middle East, North Africa vs. the Nuclear Development Option. Available at http://www.geni.org. Accessed 20 October 2007.
66. James Drummond (2009), 'Solar Power's Prospects Rise in the East', *Financial Times*, 9 November.
67. German Aerospace Center (2012), Concentrating Solar Power for the Mediterranean Region. Available at http://www.dlr.de/tt/desktopdefault.aspx/tabid-2885/4422_read-6573. Accessed 8 August 2012.

2 Morocco

1. African online news, Morocco Announces $10 billion Investment in the Energy Sector. Available at http://www.afrol.com. Accessed 9 March 2010.
2. Ministry of Energy, Mines, Water and Environment (2012) Documents. Available at http://www.mem.gov.ma. Accessed 30 June 2012.
3. Energypedia (2012) Morocco Energy Situation. Available at http://energypedia.info. Accessed 27 June 2012.
4. Renewable Energy and Energy Efficiency Partnership (2012) Policy DB Details: Morocco. Available at http://www.reeep.org. Accessed 5 July 2012.
5. Mohamed Berdai (2007) Energy Efficiency and Renewable Energy: Morocco – National Study's Summary, Centre de Developpement des Energies Renouvelables. Available at http://www.planblue.og. Accessed 30 March 2007.
6. Moroccan Agency for Solar Energy (2012) Three Challenges: Research, Training and Industrial Development. Available at http://www.masen.org. Accessed 19 April 2012.
7. Energypedia (2012) Morocco Energy Situation.
8. Afrol News (2012) Morocco Announces $10 billion Investment in the Energy Sector. Available at http://www.afrol.com. Accessed 27 June 2012.
9. MBendi Information Services (2012) Oil and Gas in Morocco – Overview. Available at http://www.mbendi.com. Accessed 27 June 2012.
10. Ibid.
11. Energy Information Administration (2012) Country Analysis Brief – Algeria. Available at http://www.eia.doe.gov. Accessed 8 March 2012.

12. Office National Des Hydrocarbures Et Des Mines (2012) History of Exploration. Available at http://www.onhym.com. Accessed 28 June 2012.
13. John R. Dyni (2012) Morocco Oil Shale. Available at http://www.geology.com. Accessed 28 June 2012.
14. Omar Bekri (2012) Possibilities for Oil Shale Development in Morocco, Center for Applied Energy Research, University of Kentucky. Available at http://www.caer.uky.edu/energeia/pdf/vol3_5.pdf. Accessed 1 July 2012.
15. Mohamed I. Shaker (2010) 'Nuclear Power in the Arab World and the Regionalization of the Nuclear Fuel Cycle: an Egyptian Perspective', *Daedalus*, 139, 93–104.
16. SMD Insights (2008) New French Nuclear Deals in the Middle East Generate Proliferation Concerns. Available at http://www.wmdinsights.com. Accessed 9 April 2008.
17. Middle East online (2012) France to help Morocco's Nuclear Energy Drive. Available at http://www.middle-east-online.com. Accessed 28 June 2012.
18. World Nuclear Association (2012) Emerging Nuclear Energy Countries. Available at http://world-nuclear.org. Accessed 2 July 2012.
19. Mohamed Berdai (2012) Energy Efficiency and Renewable Energy: Morocco – National Study's Summary. Available at http://www.planblue.org. Accessed 3 July 2012.
20. Ziinine Abdesselam (2012) Morocco: the National Renewable Energy and Efficiency Plan. Available at http://www.ajanews.asia. Accessed 3 July 2012.
21. Center for the Development of Renewable Energy (2008) Mission. Available at http://www.cder.org.ma. Accessed 20 June 2008.
22. Moroccan Investment Development Agency (2012) The Agency for Development of Renewable Energy and Energy Efficiency. Available at http://www.invest.gov.ma. Accessed 8 July 2012.
23. Power Plants around the World (2012) Hydroelectric Power Plants in Morocco. Available at http://www.industcards.com/hydro-morocco.htm. Accessed 27 June 2012.
24. Mat McDermott (2012) 1 Gigawatt Hydro-Wind Power Project Planned in Morocco Will Double Nation's Capacity. Available at http://www.treehugger.com. Accessed 27 June 2012.
25. Susan Kraemer (2012) Morocco to Pioneer 1 GW Hydro-Wind Hybrid Power. Available at http://www.greenprophet.com. Accessed 3 July 2012.
26. This section is based on Global Wind Energy Council (2012) Morocco. Available at http://www.gwec.net. Accessed 26 June 2012.
27. Eric Watkins (2012) Here Comes Morocco's Sun, Oil and Gas Journal. Available at http://www.ogj.com. Accessed 9 August 2012.
28. Amina Benkhadra (2012) The Moroccan Integrated Wind Energy Project, Ministry of Energy, Mines, Water and Environment. Available at http://www.mem.gov.ma. Accessed 8 July 2012.
29. Energypedia (2012) Eastern Africa Resource Base. Available at https://energypedia.info. Accessed 27 June 2012.
30. National Electricity Office (2012) Electrification Program. Available at http://www.one.org.ma. Accessed 4 July 2012.
31. African Development Bank Group (2012) Summary of the Environmental and Social Impact Assessment. Available at http://www.afdb.org. Accessed 21 April 2012.
32. U.S. Foreign Commercial Service (2012) Solar Energy Sector in Morocco. Available at http://www.usegyptcouncil.org. Accessed 28 June 2012.

33. Moroccan Agency for Solar Energy (2012) HM the King Presides over the Ceremony of Presentation of the Moroccan Project for Solar Energy. Available at http://www.masen.org.ma. Accessed 19 April 2012.
34. World Bank (2012) Morocco – Concentrated Solar Power Plant. Available at http://web.worldbank.org. Accessed 27 June 2012.
35. Moroccan Agency for Solar Energy (2012) Mission. Available at http://www.masen.org.ma. Accessed 4 July 2012.
36. World Association of Public Employment Services (2012) The Ouarzazate Solar Energy Complex (500 MW). Available at http://contributions.wapes.org. Accessed 21 April 2012.
37. Pilita Clark (2011) 'Morocco Wins Green Light for Solar Complex', *Financial Times*, 17 November.
38. Ansamed (2012) Morocco: EU, Nine Energy Efficiency Projects Launched Challenge Will be Responding to a New Demand by 2020. Available at http://ansamed.ansa.it. Accessed 12 April 2012.
39. Ashley Seager (2006) 'How Mirrors Can Light Up the World', *The Guardian*, 27 November.
40. EU-MED Relations (2010) The Mediterranean Solar Plan. Available at http://eu-med.blogspot.com. Accessed 10 February 2010.
41. ENP Information Center (2010) The Mediterranean Solar Plan – Working Together to Meet the Energy Challenge. Available at http://www.enpi-info.eu. Accessed 7 May 2010.
42. Enpi Information Center (2012) EU Approves Mediterranean Solar Plan Technical Assistance Project. Available at http://www.enpi-info.eu. Accessed 2 April 2012.
43. Sahara Wind Institute (2012) Background on the Sahara Wind Project. Available at http://www.saharawind.com. Accessed 5 July 2012.
44. Khalid Benhamou (2012) The Sahara Wind Project: Connecting Wind and Water. Available at http://www.saharawind.com. Accessed 5 July 2012.
45. Sahara Wind Institute (2012) Efficiency, Clean Mobility & Large Renewable Energy Projects for Germany, Europe & North Africa. Available at http://www.saharawind.com. Accessed 5 July 2012.
46. Desertec Organization (2008) The Desertec Concept. Available at http://www.desertec.org. Accessed 5 August 2008.
47. Leo Hickman (2011) Could the Desert Sun Power the World? *The Guardian*, 11 December.
48. Desertec Foundation (2012) 12 Companies Plan Establishment Of A Desertec Industrial Initiative. Available at http://www.desertec.org. Accessed 7 July 2012.
49. Desertec Foundation (2012) The Desertec Concept. Available at http://www.desertec.org/concept. Accessed 2 April 2012.
50. Norton Rose (2012) Renewable Energy in Morocco. Available at http://www.nortonrose.com. Accessed 30 May 2012.
51. Ministry of Energy (2012) Mines, Water and Environment, Documentation. Available at http://www.mem.gov.ma. Accessed 8 July 2012.
52. Deutsche Gesellschaft fur Internationale Zusammenarbeit (GIZ) GmbH (2012) Renewable Energy and Energy Efficiency Promotion in Morocco. Available at http://www.giz.de. Accessed 5 July 2012.
53. MBendi Information Services (2012) Oil and Gas in Morocco – Overview. Available at http://www.mbendi.com. Accessed 27 June 2012.

54. The Renewable Energy and Energy Efficiency Partnership (REEEP) (2012) Policy DB Details: Morocco. Available at http://www.reeep.org. Accessed 5 July 2012.
55. Norton Rose (2012) Renewable Energy in Morocco. Available at http://www.nortonrose.com. Accessed 30 May 2012.
56. Gonzalo Escribano-Frances (2012) Morocco and the European Union: Bridging Opportunities through the Mediterranean Solar Plan. An Overlook of Morocco's Perspectives. Available at http://www.sisp.it. Accessed 7 July 2012.
57. Stuart Jackson (2009) Morocco's National Plan on Solar Water Heater PROMASOL. Available at http://www.solarthermalworld.org. Accessed 10 March 2009.
58. Abdelali Dakkina (2009) Morocco's SWH Application. Available at http://www.solarthermalworld.org. Accessed 25 March 2009.

3 Egypt

1. Sohbet Karbuz and Bruno Castellano (2012) Oil and Gas, Mediterranean Energy Perspectives: Egypt. Available at http://www.ome.org. Accessed 8 April 2012.
2. Gawdat Bahgat (2009) 'North Africa and Europe: Energy Partnership', *OPEC Energy Review*, 33, 155–169.
3. British Petroleum (2011) *BP Statistical Review of World Energy*, (London: British Petroleum), p. 8.
4. Energy Information Administration (2012) Country Analysis Briefs. Available at http://www.eia.doe.gov. Accessed 20 February 2012.
5. British Petroleum (2011), BP Statistical Review of World Energy, p. 9.
6. Central Agency for Public Mobilization and Statistics (2010) *State Information Service Yearbook 2010*, (Cairo: Central Agency for Public Mobilization and Statistics), p.80.
7. British Petroleum (2011), pp. 20, 22, 23.
8. Energy Information Administration (2012).
9. Mary E. Stonaker (2010) ' Energy Infrastructure as a Diplomatic Tool: The Arab Gas Pipeline', *Journal of Energy Security*, 3. Available at http://www.ensec.org. Accessed 27 December 2010.
10. Ministry of Petroleum (2012) Ministry Strategy. Available at http://www.petroleum.gov.eg/en. Accessed 8 April 2012.
11. Egyptian General Petroleum Corporation (2012) Welcome Note. Available at http://www.egpc.com.eg. Accessed 8 April 2012.
12. Egyptian Natural Gas Holding Company (2012) About EGAS. Available at http://www.egas.com.eg. Accessed 8 April 2012.
13. Egyptian Petrochemical Holding Company (2012) About ECHEM. Available at http://www.echem-eg.com. Accessed 8 April 2012.
14. Ganoub El-Wadi Petroleum Holding Company (2012) Introduction. Available at http://www.ganope.com. Accessed 8 April 2012.
15. Egyptian Mineral Resources Authority (2012) Mission. Available at http://www.empraonline.com. Accessed 8 April 2012.
16. Sherine Nasr (2009) Liberation at a Price Tag, Al-Ahram. Available at http://weekly.ahram.org.eg. Accessed 1 April 2009.

17. Bassam Fattouh (2008) *North African Oil and Foreign Investment in Changing Market Conditions* (Oxford: Oxford Institute for Energy Studies), p. 13.
18. Franco Zallio (2010) Mediterranean Natural Gas: Greater Supply, Less Interdependence? German Marshall Fund of the United States, Policy Brief. Available at http://www.gmfus.org. Accessed 26 February 2010.
19. British Petroleum (2012) BP in Egypt. Available at http://www.bp.com. Accessed 10 April 2012.
20. ENI (2012) ENI: Agreement with Egyptian government expansion of the second LNG train in Damietta. Available at http://www.eni.gov. Accessed 10 April 2012.
21. Apache (2012) Operations. Available at http://www.apachecorp.com. Accessed 10 April 2012.
22. Shell (2012) About Shell in Egypt. Available at http://www.shell.com. Accessed 10 April 2012.
23. International Association for Natural Gas Vehicles (2012) Natural Gas Vehicle Knowledge Base. Available at http://www.iangv.org. Accessed 10 April 2012.
24. Egyptian Ministry of Petroleum (2012) Strategic Projects: Arab Gas Pipeline. Available at http://www.petroleum.gov.eg. Accessed 10 April 2012.
25. Hydrocarbons Technology (2012) SEGAS Liquefied Natural Gas Complex, Damietta, Egypt. Available at http://www.hydrocarbons-technology.com. Accessed 10 April 2012.
26. Gawdat Bahgat (2011) *Energy Security: An Interdisciplinary Approach* (London: Wiley), p.158.
27. International Energy Agency (2011) Facts on Egypt: Oil and Gas. Available at http://www.iea.org. Accessed 3 February 2011.
28. Energy Information Administration (2011) Suez Canal/SUMED Pipeline. Available at http://www.eia.doe.gov. Accessed 22 February 2011.
29. SUMED (2012) SUMED history. Available at http://www.sumed.org. Accessed 10 April 2012.
30. Federation of American Scientists (2006) Egypt: Nuclear Weapons Program. Available at http://www.fas.org. Accessed 20 February 2006.
31. Wisconsin Project on Nuclear Arms Control (2008) Egypt's Building Nuclear Program, Risk Report 2, No. 5 (September/October 1996). Available at http: //www.wisconsinproject.org/countries/egypt/nuke.html. Accessed 30 March 2008.
32. Mohamed Shaker (2010) 'Nuclear Power in the Arab World and the Regionalization of the Nuclear Fuel Cycle: An Egyptian Perspective', *Daedalus,* 139, 93–104.
33. Muhammad al-Sayed Selim (2006) Egypt and the Middle Eastern Nuclear Issue, Strategic Analysis 15, No. 1 (January 1996). Available at http://www. nti.org. Accessed 12 February 2006.
34. Shai Feldman (1997) *Nuclear Weapons and Arms Control in the Middle East* (Cambridge, MA: MIT Press), p. 59.
35. Barbara Gregory (1995) 'Egypt's Nuclear Program: Assessing Supplier-Based and other Developmental Constraints', *Nonproliferation Review,* 3, 20–27.
36. World Tribune (2001) 'Egypt Vague on Mubarak's Nuclear Deal with Russia, Early Exit from Moscow'.
37. Al-Ahram (2006) 'Egypt's Energy Minister on Nuclear Dream'.
38. WorleyParsons (2012) 1,200 Megawatt Nuclear Power Plant in Egypt. Available at http://www.worleyParsons.com. Accessed 13 April 2012.

39. Abd El-Hamid Abbas El-Desoky (2011) Building the Nuclear Power Infrastructure in Egypt. Available at http://www.iaea.org. Accessed 1 July 2011.
40. WMD Insights (2008) Algeria, Emirates Plan Nonproliferation-Friendly Nuclear Program; Egypt Keeps Fuel Cycle Options Open, Rejects Expanded IAEA Monitoring. Available at http://www.wmdinsights.com. Accessed 13 June 2008.
41. Peter Crail and Jessica Lasky-Fink (2010) Middle Eastern States Seeking Nuclear Power. Available at http://www.armscontrol.org. Accessed 27 December 2010.
42. Jessica Jewell (2011) 'A Nuclear-Powered North Africa: Just a Desert Mirage or Is There Something on the Horizon?' *Energy Policy*, 39, 4445–4457.
43. Emanuela Menichetti and Adel Soliman Beshara (2012) Renewable Energy and Energy Efficiency, pp. 283–326 in Raafat El-Beltagy (ed) Mediterranean Energy Perspectives: Egypt. Available at http://www.ome.org. Accessed 8 April 2012.
44. New and Renewable Energy Authority (2012) About NREA. Available at http://www.nrea.gov.eg. Accessed 12 April 2012.
45. Essam Hassan Ahmed (2012) Renewable Energy for Sustainable Development, Egyptian Environmental Affairs Agency. Available at http://www.eeaa.gov.eg. Accessed 2 April 2012.
46. Mohamed Shaker, 93–104.
47. John Muchira (2012) Egypt to Oppose Ethiopia's Nile Hydropower Plan. Available at http://www.engineeringnews.co.za. Accessed 2 April 2012.
48. Al-Ahram (2011) Ethiopia Hydropower Nile Project Alarms Egypt. Available at http://english.ahram.org.eg. Accessed 14 March 2011.
49. Forward Egypt (2011) Creating Wealth: Renewable Energy. Available at http: //ismahdy.wordpress.com. Accessed 5 March 2011.
50. Florian Neuhof (2011) 'Egypt Show Faith in Renewable Energy with Wind Farm Tender', *The National*, 11 November 2011.
51. Global Wind Energy Council (2012) Egypt. Available at http://www.gwec.net. Accessed 12 April 2012.
52. United Nations Environment Program (2012) Mediterranean Renewable Energy Program. Available at http://www.unep.org. Accessed 14 April 2012.
53. Solar Paces (2012) Kuraymat Integrated Solar Combined Cycle. Available at http://www.solarpaces.org. Accessed 12 April 2012.
54. Fichtnersolar (2012) Company profile. Available at http://www.fichtnersolar.com. Accessed 14 April 2012.
55. John Radcliff (2011) Kuraymat Solar Plant to Boost Egypt's Power Production, *Daily News Egypt*. Available at http://thedailynewsegypt.com. Accessed 5 July 2011.
56. International Press Service (2011) Solar Energy Projects Picking up Again after Uprising. Available at http://ipsnews.net. Accessed 22 June 2011.
57. Egyptian Cabinet of Ministers (2009) The Supreme Energy Council: The Energy Efficiency Unit. Available at http://www.jcee-eg.net. Accessed 30 October 2009.
58. Matt Bradley (2012) 'Egypt's Brewing Crisis: Subsidies', *Wall Street Journal*, 23 March.
59. British Petroleum (2011) *BP Statistical Review of World Energy*, pp. 6, 20.

4 Israel

1. Philip Hemmings (2011) Addressing Challenges in the Energy Sector in Israel. Available at http://www.oecd.org. Accessed 6 December 2011.
2. Energy Information Administration (2004) Country Analysis Brief, Israel. Available at http://www.eia.doe.gov. Accessed 8 April 2004.
3. Stephen Pierce (2004) 'Pace Speeds Up in Exploration for Triassic Targets in Israel', *Oil and Gas Journal*, 102, 42–44.
4. Zion Oil (2007) Exploration History. Available at http://www.zionoil.com /industry/exploration.html. Accessed 23 June 2007.
5. Edward Rosen (1982) 'The Effect of the Relinquished Sinai Resources on Israel's Energy Situation and Policies', *Middle East Review*, 14, 5–12.
6. Frank Kessler (1979) 'Planning Ahead: Anticipating Iranian Oil Cutoff, Israel Quietly Made Deals for Supplies from Other Sources', *Wall Street Journal*, February 9.
7. Bret Stephens (2011) 'Could Israel Become an Energy Giant?' *Wall Street Journal*, April 5.
8. Energy Information Administration (2006) Country Analysis Briefs: Eastern Mediterranean. Available at http://www.eia.doe.gov. Accessed 4 October 2006.
9. DeWayne Coxon and Aryeh Greenfield (1985) *Israel's Energy: The Search for Alternative Energy Resources in a Small Country* (Jerusalem: Israel Science Publishers), p. 62.
10. Ministry of National Infrastructures (2007) Natural Gas. Available at http: //www.mni.gov.il. Accessed 24 June 2007.
11. *Haaretz* (2007) 'Israel Eyes Liquefied Natural Gas Tender to Diversify Energy Supply', May 6.
12. Energy Information Administration (2004) Country Analysis Briefs: Israel.
13. Government of Israel (2012) Natural Gas Market Law, 5762–2002. Available at http://www.eng-archieve.antitrust.gov.il. Accessed 16 June 2012.
14. Philip Hemmings (2011) Addressing Challenges in the Energy Sector in Israel, Organization for Economic Cooperation and Development. Available at http://www.oecd.org. Accessed 20 November 2011.
15. Shmuel Even (2001) Israeli Natural Gas: the Economic and Strategic Significance, Strategic Assessment Vol. 4, No. 3. Available at http://www.tau.ac.il. Accessed 28 November 2001.
16. *Energy Economist* (2000) 'Gas and the Prospects for Middle East Peace', 226, 4–8.
17. International Energy Agency (2008) Energy Balance for Israel. Available at http://www.iea.org. Accessed 22 May 2011.
18. Ethan Bronner (2010) 'Natural Gas Deposits Improve Israel's Energy Outlook', *New York Times*, 20 August.
19. Charles Levinson and Guy Chazan (2010) 'Israel Drills Big Gas Prospect', *Wall Street Journal*, 18 October.
20. *Oil and Gas Journal (2009)* 'Tamar Wildcat Finds Subsalt Gas off Israel', 107, 22–26.
21. Avi Bar-Eli (2010) 'Netanyahu Offers Natural Gas to Greece', *Haaretz*, 29 August.
22. Charles Levinson and Guy Chazan (2010) 'Big Gas Find Sparks a Frenzy in Israel', *Wall Street Journal*, 30 December.

23. Shlomo Maital (2011) 'The Natural Gas Solution', *Jerusalem Post,* 9 May.
24. Eric Watkins (2010) 'The Scramble for East Mediterranean Gas Is On', *Oil and Gas Journal*, 108, 55–58.
25. Eric Watkins (2011) 'A Mediterranean Hotbed', *Oil and Gas Journal*, 109, 23–24.
26. *International Herald Tribune* (2005) 'Egypt and Israel Sign 15-Year Natural Gas Deal', 1 July.
27. Avi Bar-Eli (2011) 'Israel Spending Extra $1.5 million a Day after Egypt Gas Line Explosion', *Haaretz*, 11 February.
28. Eric Watkins (2011) 'Israel Takes Emergency Measures to Shore up Energy Supplies', *Oil and Gas Journal*, 109, 24–25.
29. John Steinbach (2002) Israeli Weapons of Mass Destruction: A Threat to Peace. Center for Research on Globalization. Available at http://www.globalresearch.ca. Accessed 2 March 2002.
30. Yair Evron (1979) Israel and nuclear weapons, in *Nuclear Proliferation in Developing Countries*, ed. Jae Kyu Park, 123–141 (Seoul: Institute for Far Eastern Studies).
31. Wisconsin Project on Nuclear Arms Control (2002) Israel's Nuclear Weapon Capability, An overview, Risk Report, Vol. 2, No. 4, July/August 1996. Available at http://www.wisconsinproject.org. Accessed 1 August 2002.
32. Michell Reiss (1988) *Without the Bomb: The Politics of Nuclear Nonproliferation* (New York: Columbia University Press), p. 149.
33. Federation of American Scientists (2002) Nuclear Weapons. Available at http://fas.org/nuke/guide/israel/nuke. Accessed 20 October 2002.
34. Yossi Melman (2011) 'Israel to Stress Safety of Its Nuclear Reactors at IAEA special session', *Haaretz*, 20 June.
35. Israel Atomic Energy Commission (2012) Nuclear Research Center Negev – NRCN. Available at http://iaec.gov.il. Accessed 17 June 2012.
36. Israel Atomic Energy Commission (2012) Nuclear Research Center – Soreq. Available at http://iaec.gov.il. Accessed 17 June 2012.
37. Israel Atomic Energy Commission (2012) Israel's Activity at the IAEA. Available at http://iaec.gov.il. Accessed 17 June 2012.
38. Israel Atomic Energy Commission (2012) About us. Available at http://iaec.gov.il. Accessed 12 June 2012.
39. Prime Minister's Office (2012) Atomic Energy Commission. Available at http://www.pmo.gov.il. Accessed 18 June 2012.
40. Steven Erlanger (2010) 'Israel Intends to Build Civilian Nuclear Plants', *New York Times*, 9 March.
41. Amiram Barkat (2011) Israel Will Not Have a Nuclear Power Plant in Next 20–30 years, Israel's Business Arena. Available at http://www.globes.co.il. Accessed 8 November 2011.
42. Barak Ravid and Zafrir Rinat (2011) 'PM: Israel Reconsidering Nuclear Power Plans in Light of Japan Crisis', *Haaretz*, 17 March.
43. Arava Power Company (2012) The Vision of Ben-Gurion. Available at http://www.aravapower.com. Accessed 22 May 2012.
44. Itay Zetelny (2012) Renewable Energy Recap: Israel, Ernst and Young. Available at http://www.renewableenergyworld.com. Accessed 22 May 2012.
45. European Union (2012) Paving the way for the Mediterranean Solar Plan. Available at http://www.pavingtheway-msp.eu. Accessed 20 June 2012.

46. Ministry of Industry (2012) Trade and Labor, National Sustainable Energy and Water Program. Available at http://israelnewtech.gov.il. Accessed 16 May 2012.
47. Ernst & Young (2011) Policies to Support Eco-innovation in Israel. Available at http://www.oecd.org. Accessed 20 December 2011.
48. Renewable Energy Association of Israel (2012) Welcome to the Renewable Energy Association of Israel. Available at http://www.renewable.org.il. Accessed 1 June 2012.
49. Israel Export and International Cooperation Institute (2012) Israel's Renewable Energy Overview. Available at http://www.slideshare.net. Accessed 18 May 2012.
50. Arava Power Company (2012) Company Profile. Available at http://www.aravapower.com. Accessed 22 May 2012.
51. Arava Power Company (2012) Ketura Sun Inauguration. Available at http://www.aravapower.com. Accessed 22 May 2012.
52. Isabel Kershner (2012) 'Israeli Desert Yields a Harvest of Energy', *New York Times*, 20 April.
53. Itay Zetelny (2012) Renewable Energy Recap: Israel, Ernst and Young. Available at http://www.renewableenergyworld.com. Accessed 22 May 2012.
54. Israel Wind Power (2012) Israel Wind Power. Available at http://www.israel-windpower.com. Accessed 22 May 2012.
55. Jewish Virtual Library (2012) Israel Energy Information. Available at http://www.jewishvirtuallibrary.org. Accessed 23 June 2012.
56. Ministry of National Infrastructure (2012) About Us. Available at http://www.mni.gov.il. Accessed 23 June 2012.
57. Public Utilities Authority (2012) About Us. Available at http://www.pua.gov.il. Accessed 23 June 2012.
58. Michael Ratner (2011) Israel's Offshore Natural Gas Discoveries Enhance its Economic and Energy Outlook, Congressional Research Service. Available at http://www.crs.gov. Accessed 31 January 2011.
59. BIRD (2012) What is BIRD. available at http://www.bird.com. Accessed 1 June 2012.
60. BIRD Energy (2012) What is BIRD Energy? Available at http://www.birdf.com. Accessed 1 June 2012.

5 Saudi Arabia

1. Ottma Edenhofer, Ramon Pichs Madruga, Youba Sokona (2012) *Renewable Energy Sources and Climate Change Mitigation*, (Cambridge: Cambridge University Press), p. 164.
2. ExxonMobil (2012) 2012 The Outlook for Energy: A View to 2040. Available at http://www.exxonmibil.com. Accessed 8 June 2012.
3. International Energy Agency (2011) The IEA Model of Short-term Energy Security: Primary Energy Sources and Secondary Fuels. Available at http://www.iea.org. Accessed 10 December 2011.
4. Unless stated otherwise, all figures are from British Petroleum (2011) BP Statistical Review of World Energy, London.

5. Zawya Dow Jones (2010) Saudi Establishes New City for Nuclear Energy Research. Available at http://www.zawya.com. Accessed 17 April 2010.
6. International Monetary Fund (2011) Regional Economic Outlook: Middle East and Central Asia, Washington DC, p. 1.
7. Peter Meisen and Lesley Hunter (2007) Renewable Energy Potential of the Middle East and North Africa vs. the Nuclear Development Option, Global Energy Network Institute. Available at http://www.geni.org. Accessed 20 October 2007.
8. Saudi Arabia Embassy in Washington DC (2012) Water Resources. Available at http://www.saudiembassy.net. Accessed 17 January 2012.
9. Abber Allam (2011) 'Saudi to Develop Solar and Nuclear Power,' *Financial Times*, 24 January.
10. ABB Group (2012) Saudi Arabia Energy Efficiency Report. Available at http://www05.abb.com/global/scot/scot316.nsf/ ... /saudi%20arabia.pdf. Accessed 18 January 2012.
11. Arabian Business (2010) Top Oil Exporter Saudi Arabia and France Are Preparing to Sign an Agreement on Cooperation in Civilian Nuclear Energy. Available at http://www.Arabianbusiness.com, Accessed 20 December 2010.
12. Angus McDowall (2011) 'Rising Saudi Thirst for Oil Drives Plans to Go Nuclear', *Wall Street Journal*, 23 June.
13. Sarah Mishkin (2011) 'Domestic Oil Usage to Vie with Exports', *Financial Times*, 9 May.
14. Eric Watkins (2011) 'Saudi Arabia Going Nuclear', *Oil and Gas Journal*, (109) 26–28.
15. Abber Allam (2011) 'Saudi to Develop Solar and nuclear Power', *Financial Times*, 24 January.
16. Javier Blas and Guy Chazan (2012) 'Saudi Arabia Targets $100 Crude Price', *Financial Times*, 16 January.
17. World Nuclear Association (2012) Nuclear Power in the World Today. Available at http://www.world-nuclear.org. Accessed 8 June 2012.
18. Mohamed I. Shaker (2010) 'Nuclear Power in the Arab World and the Regionalization of the Nuclear Fuel Cycle: An Egyptian Perspective', *Doedalus*, 139, 93–104.
19. Associated Press (2006) 'Arab Nations Urged to Enter Nuclear Club'. Available at http://www.nytimes.com. Accessed 28 March 2006.
20. Roula Khalaf (2006) 'Gulf Arabs Weigh Joint Nuclear Program', *Financial Times*, 10 December.
21. Scotsman (2011) Saudi Arabia Announces Details of Nuclear Power Boom. Available at http://www.scotsman.com. Accessed 1 June 2011.
22. Tamsin Carlisle (2011) 'Saudi Arabia in Agreement to Explore Nuclear Power', *The National*, 23 February.
23. Al-Arabiya News (2011) Saudi Arabia Signs Nuclear Energy Deal with Argentina. Available at http://www.alarabiya.net. Accessed 29 June 2011.
24. Reuters (2011) Saudi Arabia, South Korea in Nuclear Cooperation Deal. Available at http://www.reuters.com. Accessed 15 November 2011.
25. Summer Said (2012) 'Saudi Arabia, China Sign Nuclear Cooperation Pact', *Wall Street Journal*, 16 January.
26. Patrick W. Ryan (2011) Bilateral Civilian Nuclear Trade Talks Set to Open, Saudi-US Relations Information Service. Available at http://www.susris.com. Accessed 29 July 2011.

27. The Energy Collective (2011) Saudi Arabia's Nuclear Energy Ambitions. Available at http://www.theenergycollective.com. Accessed 23 August 2011.
28. Paul K. Kerr and Mary Beth Nikitin (2010) Nuclear Cooperation with Other Countries: A Primer, Congressional Research Service. Available at http: //www.crs.gov. Accessed 20 June 2010.
29. Elaine M. Grossman (2011) Obama Team Eyes Saudi Nuclear Trade Deal without Nonproliferation Terms, Global Security Newswire. Available at http://gsn.nti.org. Accessed 25 January 2011.
30. Department of State (2012) Global Initiative to Combat Terrorism. Available at http://www.state.gov. Accessed 22 January 2012.
31. Department of State (2012) Proliferation Security Initiative. Available at http://www.state.gov. Accessed 22 January 2012.
32. International Atomic Energy Agency (2012) Factsheets and FAQs: IAEA Safeguards Overview. Available at http://www.iaea.org. Accessed 22 January 2012.
33. Ottma Edenhofer, Ramon Pichs Madruga, and Youba Sokona (2012) *Renewable Energy Sources and Climate Change Mitigation* (Cambridge: Cambridge University Press), p. 178.
34. British Petroleum (2012) Energy Outlook 2030. Available at http://www. bp.com. Accessed 30 January 2012.
35. Peter Meisen and Lesley Hunter (2007) Renewable Energy Potential of the Middle East and North Africa vs. the Nuclear Development Option, Global Energy Network Institute. Available at http://www.geni.org. Accessed 20 October 2007.
36. Judith Miller (1983) 'In Saudi Arabia, the Sun Shines Bright on Solar Power', *New York Times*, 1 November.
37. Justin Dargin (2009) 'Saudi Arabia, UAE Promote Energy From Sun and Wind', *Oil and Gas Journal*, (107) 18–22.
38. Eric Watkins (2011) 'Saudis Go Solar', *Oil and Gas Journal*, 109, 24–26.
39. Anthony DiPaola (2011) Saudi Arabia to Target Solar Power in $100 Billion Energy Plan, Bloomberg. Available at http://www.bloomberg.com. Accessed 31 March 2011.
40. Ahmed al-Zila'I (2012) Saudi Arabia's First Solar Power Station Inaugurated on Farasan Island, Al-Arabiya. Available at http://www.alarabiya.net. Accessed 17 January 2012.
41. Solar Frontier (2012) Saudi Aramco North Park. Available at http://www. solar-frontier.com. Accessed 29 January 2012.
42. P.K. Abdul Ghafour (2011) Saudi Arabia to Exploit Solar Energy for Desalination Plants, Arab News. Available at http://arabnews.com. Accessed 2 December 2011.
43. April Yee (2011) 'Saudi Arabia Embraces Solar Energy', *The National*, 10 March.
44. Anthony DiPaola (2011) Saudi Arabia to Target Solar Power in $100 Billion Energy Plan, Bloomberg. Available at http://www.bloomberg.com. Accessed 31 March 2011.
45. National Solar Systems (2012) Providing Clean Energy from Saudi Arabia to the World. Available at http://www.national-solar.net. Accessed 17 January 2012.
46. Javid R. Laghari (2012) Empires of the Mind, Newsweek Pakistan. Available at http://www.newsweekpakistan.com. Accessed 3 February 2012.

47. Government of Saudi Arabia (2012) Royal Decree Establishing King Abdullah City for Atomic and Renewable Energy. Available at http://www.energy.gov.sa. Accessed 29 January 2012.
48. Government of Saudi Arabia (2012) A New Era of Sustainable Energy. Available at http://www.energy.gov.sa. Accessed 29 January 2012.
49. King Abdullah Petroleum Studies and Research Center (2012) About us. Available at http://www.kapsarc.org. Accessed 8 June 2012.
50. Saudi Aramco (2012) CEO Sets Ambitious Plan to Train 2 Million Saudis. Available at http://www.Saudiaramco.com. Accessed 1 June 2012.
51. Energy Information Administration (2012) Energy Efficiency. Available at http://www.eia.doe.gov. Accessed 31 January 2012.
52. ABB Group (2012) Saudi Arabia Energy Efficiency Report. Available at http://www05.abb.com. Accessed 31 January 2012.
53. Glada Lahn and Paul Stevens (2011) Burning Oil to Keep Cool: The Hidden Energy Crisis in Saudi Arabia, Chatham House. Available at http://www.chathamhouse.org. Accessed 30 December 2011.
54. International Energy Agency (2010) Analysis of the Scope of Energy Subsidies and Suggestions for the G-20 Initiative. Available at http://www.iea.org. Accessed 16 June 2010.
55. Yousef Alyousef and Paul Stevens (2011) 'The Cost of Domestic Energy Prices to Saudi Arabia', *Energy Policy*, 39, 6900–6905.
56. United Nations Development Program (2011) National Energy Efficiency Program. Available at http://www.undp.org.sa. Accessed 17 January 2011.
57. King Abd al-Aziz City for Science and Technology (2012) Saudi Center for Energy Efficiency. Available at http://www.kacst.edu.sa. Accessed 31 January 2012.
58. British Petroleum (2012) BP Energy Outlook 2030. Available at http://www.bp.com. Accessed 30 January 2012.

6 The United Arab Emirates

1. Jessica Jewell (2011) The IEA Model of Short-term Energy Security (MOSES): Primary Energy Sources and Secondary Fuels, International Energy Agency. Available at http://www.iea.org. Accessed 10 December 2011.
2. International Monetary Fund (2011) *Regional Economic Outlook: Middle East and Central Asia* (Washington, DC: International Monetary Fund), p. 83.
3. British Petroleum (2011) *BP Statistical Review of World Energy*, (London), p. 6.
4. UAE Interact (2012) Pearls to Petroleum Transforms Abu Dhabi. Available at http://www.uaeinteract.com. Accessed 22 March 2012.
5. Abu Dhabi National Oil Company (2012) About ADNOC. Available at http://www.adnoc.ae. Accessed 25 March 2012.
6. Abu Dhabi National Oil Company (2012) The Supreme Petroleum Council. Available at http://www.adnoc.ae/content. Accessed 25 March 2012.
7. Abu Dhabi National Oil Company (2012) ADNOC Technical Institute. Available at http://www.adnoc.ae. Accessed 25 March 2012.
8. Petroleum Institute (2012) Brief History of the Petroleum Institute. Available at http://www.pi.ae. Accessed 22 March 2012.

9. Abu Dhabi Company for Onshore Oil Operations (2012) Overview. Available at http://www.adco.ae. Accessed 25 March 2012.
10. Abu Dhabi Marine Operating Company (2012) About ADMA-OPCO. Available at http://www.adma-opco.com. Accessed 25 March 2012.
11. National Gas Shipping Company (2012) About us. Available at http://www.ngsco.com/en/adnatco/about/index.html. Accessed 25 March 2012.
12. United Arab Emirates Embassy, Washington DC (2012) The UAE and Global Oil Supply. Available at http://www.uae.embassy.org. Accessed 22 March 2012.
13. Javier Blas (2012) Pipelines Bypassing Hormuz Open, *Financial Times*, 15 July.
14. Energy Information Administration (2011) Country Analysis Briefs: United Arab Emirates. Available at http://www.eia.doe.gov. Accessed 21 January 2011.
15. Abu Dhabi Gas Industries (2012) Our story. Available at www.gasco.ae. Accessed 28 March 2012.
16. Hydrocarbons Technology (2012) CASCO Integrated Gas Development Project. Available at http://www.hydrocarbons-technology.com. Accessed 28 March 2012.
17. National Gas Shipping Company (2012) About us. Available at http://www.ngsco.com. Accessed 28 March 2012.
18. British Petroleum, p. 20.
19. UAE Interact (2010) Population Leaps to 8.19 million. Available at http://www.uaeinteract.com. Accessed 30 May 2010.
20. Stroytransgaz (2012) Abu Dhabi – Gas Pipeline and Dolphin Project. Available at http://www.stroytransgaz.com. Accessed 28 March 2012.
21. Tamsin Carlisle (2010) Iran Cancels Gas Deal with Crescent Petroleum. Available at http://www.thenational.ae. Accessed 7 August 2010.
22. Justin Dargin (2012) Policy Brief: The Dubai Initiative. Available at http://belfercenter.ksg.harvard.edu. Accessed 29 March 2012.
23. British Petroleum, p. 9.
24. Abdullah bin Zazed Al-Nahyan (2008) 'Why Go Nuclear?' *Bulletin of the Atomic Scientists*, 64, 16–17.
25. Raid Qusti (2006) 'GCC to Develop Civilian Nuclear Energy,' *Arab News*, 11 December.
26. United Arab Emirates, Policy of the United Arab Emirates on the Evaluation and Potential Development of Peaceful Nuclear Energy (2011) Available at http://www.carnegieendowment.org/static/npp/reports/UAE_white_paper.pdf. Accessed 29 December 2011.
27. Federal Authority for Nuclear Regulation (2011) Mission. Available at http://www.fanr.gov.ae. Accessed 29 December 2011.
28. Emirates Nuclear Energy Corporation (2011) About us. Available at http://www.enec.gov.ae. Accessed 29 December 2011.
29. International Advisory Board (2011) About us. Available at http://www.uaeiab.ae. Accessed 29 December 2011.
30. Christopher M. Blanchard and Paul K. Kerr (2009) The United Arab Emirates Nuclear Program and Proposed U.S. Nuclear Cooperation, Congressional Research Service. Available at http://www.crs.gov, Accessed 11 June 2009.
31. Nuclear Engineering International (2011) First Steps of a Brand-New Regulator. Available at http://www.neimagazine.com. Accessed 28 December 2011.

32. Embassy of the United Arab Emirates in Washington, DC (2010) US-UAE 123 Agreement Enters into Force. Available at http://www.UAE-embassy.org/media/press-releases/17-Dec-2009. Accessed 18 October 2010.
33. Fred McGoldrick (2010) The US-UAE Peaceful Nuclear Cooperation Agreement: A Gold Standard or Fool's Gold? Center for Strategic and International Studies. Available at http://www.csis.org, Accessed 30 November 2010.
34. Ellen Tauscher (2011) Agreement for Cooperation between the Government of the United States of America and the Government of the United Arab Emirates Concerning Peaceful Uses of Nuclear Energy, U.S. Department of State. Available at http://www.state.gov. Accessed 24 January 2011.
35. Robin Wigglesworth (2009) 'UAE and US Sign Nuclear Deal,' *Financial Times*, 18 January.
36. Paul K. Kerr and Mary Beth Nikitin (2010) Nuclear Cooperation with other Countries: A Primer, Congressional Research Service. Available at http://www.crs.gov. Accessed 28 May 2010.
37. Christopher M. Blanchard and Paul K. Kerr (2009) The United Arab Emirates Nuclear Program and Proposed US Nuclear Cooperation.
38. Business Week (2011) UAE's Nuclear Power Program Said to Cost $30 Billion. Available at http://www.businessweek.com. Accessed 28 December 2011.
39. Hamad Al Kaabi (2011) 'United Arab Emirates and the Experience of a Nuclear Newcomer,' *Bulletin of the Atomic Scientists*, 67, 52–59.
40. Peter Shaw-Smith (2010) 'High Stakes Played for in UAE Nuclear Scheme', *Financial Times*, 18 October.
41. World Nuclear Association (2012) Nuclear Power in the United Arab Emirates. Available at http://world-nuclear.org/info/UAE_nuclear_power_inf123.html. Accessed 6 July 2012.
42. Federal Authority for Nuclear Regulation (2010) Training Program: Gulf Nuclear Energy Infrastructure Institute (GNEII). Available at http://fanr.gov.ae. Accessed 27 December 2010.
43. Andrew England (2007) 'Abu Dhabi Eyes Renewable Energy Future,' *Financial Times*, 4 April.
44. James Drummond and Robin Wigglesworth (2009) 'Renewables Race Hots Up', *Financial Times*, 22 January.
45. Emirates Nuclear Energy Corporation (2011) The Case for Nuclear Energy. Available at http://www.enec.gov.ae. Accessed 1 January 2011.
46. Justin Dargin (2009) 'Saudi Arabia, UAE Promote Energy from Sun and Wind', *Oil and Gas Journal*, 107, 18–22.
47. British Broadcasting Service (2008) Work Starts on Gulf Green City. Available at http://newsvote.bbc.co.uk. Accessed 10 February 2008.
48. Masdar (2011) About Us. Available at http://www.masdar.ae. Accessed 24 January 2011.
49. *Wall Street Journal* (2010) Abu Dhabi Picks Partners for Solar Project, 10 June.
50. Total (2011) Masdar Partners with Total and Abengoa Solar to Build the World's Largest Concentrated Solar Power Plant. Available at http://www.total.com. Accessed 24 January 2011.
51. Carola Hoyos and Robin Wigglesworth (2010) 'Total and Abengoa to Sign UAE Solar Farm Deal', *Financial Times*, 8 June.
52. Arabian Business (2011) Masdar's Shams 1 Solar Plant to Start by Q3 2012, 14 November.

53. AME Info FZ LLC (2008) Masdar and Hydrogen Energy Plan Clean Energy Plant in Abu Dhabi. Available at http://www.ameinfo.com. Accessed 22 January 2008.
54. April Yee (2011) 'BP Delays $2 Billion Hydrogen Power Plant in Abu Dhabi', *The National*, 19 January.
55. Smart Power Market Watch (2011) Masdar (Abu Dhabi Net Zero Carbon City) Starts Geothermal Drilling. Available at http://www.smartpowercommunity.com. Accessed 24 January 2011.
56. Chris Stanton (2009) 'Drilling Deep for Green Energy', *The National*, 28 September.
57. Andrew England (2007) 'Abu Dhabi Eyes Renewable Energy Future', *Financial Times*, 4 April.
58. Masdar Institute of Science and Technology (2011) Vision. Available at http://www.masdar.ac.ae. Accessed 31 December 2011.
59. Masdar Institute of Science and Technology (2011) Mission. Available at http://www.masdar.ac.ae. Accessed 31 December 2011.
60. International Energy Agency (2010) Analysis of the Scope of Energy Subsidies and Suggestions for the G-20 Initiative. Available at http://www.iea.org. Accessed 16 June 2010.

7 Iran

1. United States Department of State (2012) Treaty of the Non-Proliferation of Nuclear Weapons. Available on line at http://www.state.gov. Accessed 4 March 2012.
2. British Petroleum (2011) *BP Statistical Review of World Energy*, London, pp. 6, 20.
3. Kamal Kharrazi (2005) 'The View from Tehran', *Middle East Policy*, 12, 25–30.
4. Thomas W. Wood, Matthew D. Milazzo, Barbara A. Reichmuth, and Jeffrey Bedell (2007), 'The Economics of Energy Independence for Iran', *Non-Proliferation Review*, 14, 89–112.
5. Wyn Q. Bowen and Joanna Kidd (2004) 'The Iranian Nuclear Challenge,' *International Affairs*, 80, 257–276.
6. Energy Information Administration (2012) Country Analysis Briefs: Iran. Available on line at http://www.eia.doe.gov. Accessed 17 February 2012.
7. National Iranian Oil Company (2009) Brief History of Iran Oil Nationalization. Available at http://www.nioc.ir/English/brief_history/index.html. Accessed 13 September 2009.
8. George Lenczowski (1960) *Oil and State in the Middle East* (Ithaca, NY: Cornell University Press), p. 124.
9. Keith McLachlan (1960) 'Oil in the Persian Gulf Area', in *The Persian Gulf States: A General Survey* (ed. Alvin Cottrell), (Baltimore, MD: Johns Hopkins University Press), pp. 195–224.
10. George Lenczowski (1960) *Oil and State in the Middle East* (Ithaca, NY: Cornell University Press), p. 126.
11. Daniel Brumberg and Ariel Ahram (2007) *The National Iranian Oil Company in Iranian Politics* (Houston, TX: James A. Baker III Institute for Public Policy at Rice University), p. 4.
12. Valerie Marcel and John Mitchell (2006), *Oil Titans: National Oil Companies in the Middle East* (Washington, DC: Brookings Institution), p. 42.

13. Plran Shammas (2001) 'Iran: Review of Petroleum Developments and Assessments of the Oil and Gas Fields'. *Energy Exploration and Exploitation'*, 19, 207–260.
14. Narsi Ghorban (2006) 'Monetizing Iran's Gas Resources and the Debate over Gas Export and Gas-based Industries Options', *Middle East Economic Survey*, 49, 26–28.
15. British Petroleum (2011), *BP Statistical Review of World Energy*, London, pp. 20, 22.
16. Manoochehr Takin (1999) 'Iranian Gas to Europe?' *Middle East Economic Survey*, 42, 18–20.
17. National Iranian Gas Company (2012) Iran's Gas Industry History. Available on line at http://www.nigc.ir. Accessed 27 February 2012.
18. Samsam Bakhtiari and Farzad Shahbudaghlou (2000) 'Energy Consumption in the Islamic Republic of Iran', *OPEC Review*, 24, 211–233.
19. Energy Information Administration (2012) Country Analysis Briefs, Iran. Available on line at http://www.eia.doe.gov. Accessed 17 February 2012.
20. National Iranian Gas Company (2012) NIGC Strategic Objectives. Available on line at http://www.nigc.com.ir. Accessed 27 February 2012.
21. Pars Special Economic Energy Zone (2012) South Pars Gas Field. Available on line at http://www.pseez.ir. Accessed 27 February 2012.
22. Pars Oil and Gas Company (2009) About POGC. Available on line at http: //www.pogc.ir. Accessed 13 September 2009.
23. Pars Oil and Gas Company (2009) South Pars Gas Field. Available at http: //pogc.ir. Accessed 13 September 2009.
24. Dean Gaddy (2001) 'Iran Expands Middle East Influence', *Oil & Gas Journal*, 99, 74–81.
25. Ian Seymour (2000) 'Kuwait's Upstream Oil Opening in the Context of Parallel Development Elsewhere', *Middle East Economic Survey*, 43, 24–26.
26. Kirsten Bindemann (1999) *A Little Bit of Opening Up: The Middle East Invites Bids by Foreign Oil Companies* (Oxford: Oxford Institute for Energy Studies), p. 6.
27. Fereidun Fesharaki and Mohamed Varzi (2000) 'Investment Opportunities Starting to Open up in Iran's Petroleum Sector', *Oil and Gas Journal*, 98, 44–52.
28. Alexander Brexendorff and Christian Ule (2004) 'Changes Bring New Attention to Iranian Buyback Contracts', *Oil and Gas Journal*, 102, 16–21.
29. Madjid Abbaspour (2005) Climate Policy and Sustainable Development: Opportunities for Iranian-German Cooperation. Available at http://www.ceers.org. Accessed 30 May 2005.
30. Farshad Lavizeh (2012) General Report on Geothermal and Related Fields in Iran. Available at http://kgvrs.mine.kyushu-u.ac.jp. Accessed 6 March 2012.
31. Hossein Yousefi, Sachio Ehara and Younes Noorollahi (2012) Geothermal Potential Site Selection using GIS in Iran. Available at http://www.geospatial-world.net. Accessed 10 March 2012.
32. Farideh Atabi (2004) 'Renewable Energy in Iran: Challenges and Opportunities for Sustainable Development', *International Journal of Environmental Science & Technology*, 1, 69–80.
33. Younes Noorollahi, Hossein Yousefi, Sachio Ehara, and Ryuichi Itoi (2009) Geothermal Energy Development in Iran, *Renewable and Sustainable Energy Reviews*, 13, 1127–1132.

34. Payvand (2011) Iran to Build First 5-megawatt Geothermal Plant. Available at http://www.payvand.com. Accessed 28 August 2011.
35. Hossein Yousefi Sachio Ehara, Amin Yousefi, and Fariba Seiedi (2009) Environmental Impact Assessment of Sabalan Geothermal Power Plant, NW Iran, Proceedings, Thirty-Fourth Workshop on Geothermal Reservoir Engineering (Stanford, CA: Stanford University), p. 2.
36. Global Wind Energy Council (2012), Iran: Total Installed Capacity. Available at http://www.gwec.net. Accessed 5 March 2012.
37. Fars News Agency, (2011) Iran Inaugurates First Wind-Solar Hybrid Power Plant. Available at http://english.farsnews.com. Accessed 1 July 2011.
38. Peter Meisen and Lesley Hunter (2007) Renewable Energy Potential of the Middle East, North Africa vs. the Nuclear Development Option. Available at http://www.geni.org. Accessed 20 October 2007.
39. Fars News Agency (2011) Iran to Develop Solar Energy to Decrease Fossil Fuel Consumption. Available at http://english.farsnews.com. Accessed 23 January 2011.
40. Gavin Hudson (2009) First Solar Energy Plant Completed in Iran. Available at http://ecolocalizer.com. Accessed 2 January 2009.
41. Fars News Agency (2011) Iran Opens Major Solar Power Plant. Available at http://english.farsnews.com. Accessed 22 July 2011.
42. Renewable Energy Organization of Iran (2012) History and Objectives of Iran Renewable Energy Organization (SUNA). Available at http://www.suna.org.ir. Accessed 5 March 2012.
43. Mark Fitzpatrick (2008) 'Will Nuclear Energy Plans in the Middle East Become Nuclear Weapons Strategies?' *International Relations*, 22, 381–385.
44. Peter Crail & Jessica Lasky-Fink (2008) Middle Eastern States Seeking Nuclear Power, Arms Control Association. Available at http://www.armscontrol.org. Accessed 6 May 2008.
45. Greg Bruno (2008) Iran's Nuclear Program, Council on Foreign Relations. Available at http://www.cfr.og. Accessed 5 September 2008.
46. Mohamed Ghannadi-Maragheh (2002) Atomic Energy Organization of Iran. Available at http://www.world-nuclear.org. Accessed 6 September 2002.
47. Mohamed Sahimi (2004) Iran's Nuclear Energy Program: From the United States Offering Iran Uranium Enrichment Technology to Suggestions for Creating Catastrophic Industrial Failure. Available at http://www.payvand.com. Accessed 22 December 2004.
48. Nuclear Threat Initiative (2005) Iran: Nuclear Overview. Available at http://www.nti.org. Accessed 30 October 2005.
49. Mohamed Sahimi (2003) Iran's Nuclear Program: Its History. Available at http://www.payvand.com. Accessed 3 October 2003.
50. Frank Hippel (2008) National Fuel Stockpiles: An Alternative to a Proliferation of National Enrichment Plants? Available at http://www.armscontrol.org. Accessed 3 September 2008.
51. Oliver Meier (2006) Iran and Foreign Enrichment: a Troubled Model. Available at http://www.armscontrol.org. Accessed 14 January 2006.
52. Mustafa Kibaroglu (2006) 'Good for the Shah Banned for the Mullahs: the West and Iran's Quest for Nuclear Power', *Middle East Journal*, 60, 207–233.
53. Mustafa Kibaroglu (2007) 'Iran's Nuclear Ambitions from a Historical Perspective and the Attitude of the West', *Middle Eastern Studies*, 43, 223–245.

54. Mark Fitzpatrick (2007) 'Can Iran's Nuclear Capability Be Kept Latent?' *Survival*, 49, 33–58.
55. Abbas Maleki (2010) 'Iran's Nuclear File: Recommendations for the Future', *Daedalus*, 139, 105–116.
56. Thomas Wood, Matthew Milazzo, Barbara Reichmuth, and Jeffrey Bedell. (2007) 'The Economics of Energy Independence for Iran', *Nonproliferation Review*, 14, 89–112.
57. David Albright and Jacqueline Shire (2007) A Witches' Brew? Evaluating Iran's Uranium-Enrichment Progress. Available at http://www.armscontrol.org. Accessed 3 November 2007.
58. Iran Energy Efficiency Organization (2012) Historical Background of Organization. Available at http://www.saba.org.ir. Accessed 5 March 2012.
59. Iranian Fuel Conservation Organization (2012) About Us. available at http://www.ifco.ir. Accessed 18 March 2012.

8 Conclusion

1. Richard Lester and Robert Rosner (2009) 'The Growth of Nuclear Power: Drivers and Constraints', *Daedalus*, 138, 19–30.
2. World Bank (2009) Tapping a Hidden Resource: Sustainable Development Network Middle East and North Africa Region. Available at http://www.worldbank.org. Accessed 10 February 2009.
3. Richard Ward (2010) *Sustainable Energy Security* (London: Chatham House), p.38.
4. Jessica Jewell (2011) The IEA Model of Short-term Energy Security (MOSES), International Energy Agency. Available at http://www.iea.org. Accessed 10 December 2011.
5. Andreas Goldthau and Benjamin K. Sovacool (2011), 'The Uniqueness of the Energy Security, Justice, and Governance Problem', *Energy Policy*, 41, 232–240.
6. International Energy Agency (2012) World Energy Outlook. Available at http://www.iea.org. Accessed 21 July 2012.
7. International Energy Agency (2010) Analysis of the Scope of Energy Subsidies and Suggestions for the G-20 Initiative. Available at http://www.iea.org. Accessed 16 June 2010.
8. Yousef Alyousef and Paul Stevens (2011) 'The Cost of Domestic Energy Prices to Saudi Arabia', *Energy Policy*, 39, 6900–6905.
9. Cited in Benedict Clements, Rejane Hugouneng, and Gerd Schwartz. (1995) *Government Subsidies: Concepts, International Trends and Reform Options*, IMF Working Paper, September 95/91, (Washington, DC: International Monetary Fund), p. 5.
10. Uruguay Round Agreement (2012) Agreement on Subsidies and Countervailing Measures. Available at http://www.wto.org. Accessed 21 July 2012.
11. Glada Lahn and Paul Stevens (2011) *Burning Oil to Keep Cool: the Hidden Energy Crisis in Saudi Arabia* (London: Chatham House), p. 14.
12. Bassam Fattouh and Laura El-Katiri (2012) Energy Subsidies in the Arab World, Arab Human Development Report. Available at http://www.arab-hdr.org. Accessed 2 April 2012.

13. Jeffrey Ball (2012) 'Tough Love for Renewable Energy', *Foreign Affairs*, 91, 36–44.
14. International Energy Agency (2012) Medium-term Renewable Energy Market Report 2012. Available at http://www.iea.org. Accessed 6 July 2012.
15. Dominique Guillaume, Roman Zytek, and Mohammad Reza Farzin (2011) *Iran – the Chronicles of the Subsidy Reform*. IMF Working Paper 11/167 (Washington, DC: International Monetary Fund), p. 6.
16. International Monetary Fund (2011) IMF Executive Board Concludes 2011 Article IV Consultation with the Islamic Republic of Iran. Available at http://www.imf.org. Accessed 3 August 2011.
17. International Energy Agency (2012) Energy Security is Best Achieved through Efficiency, Diversity and Cooperation'. Available at http://www.iea.org. Accessed 6 July 2012.
18. ABB Group (2012) The State of Global Energy Efficiency. Available at http://www.abb.com. Accessed 24 July 2012.
19. British Petroleum (2012) *BP Energy Outlook 2030*, London, p. 55.
20. World Bank (2009), Tapping a Hidden Resource: Sustainable Development Network Middle East and North Africa Region.
21. International Energy Agency(2012) *Energy Technology Perspectives*, Paris, p. 1.
22. International Energy Agency (2010) *Energy Efficiency Governance Handbook*, Paris, p. 7.
23. International Energy Agency (2012) Medium-Term Renewable Energy Market Report, Paris, p.11.
24. Peter Meisen (2010) Cross-Border Interconnections on Every Continent, Global Energy Network Institute. Available at http://www.geni.org. Accessed 30 June 2010.
25. European Union (2012) Seven Countries Interconnection Project. Available at www.medemip.eu. Accessed 26 July 2012.
26. Power Engineering International (2012) Transmission Grid Network: the Seven Wonders of Interconnection. Available at http://www.powerengineeringint.com. Accessed 2 April 2012.
27. Hassan K. Al-Asaad and Adnan I. Al-Mohaisen (2012) GCC Power Grid: Transforming the GCC Power Sector into a Major Energy Trading Market. Available at http://www.gccia.com.sa. Accessed 21 April 2012.
28. Hassan K. Al-Asaad and Ahmed A. Ebrahim (2012) The GCC Power Grid: Benefits and Beyond. Available at http://www.gccia.com.sa. Accessed 21 April 2012.
29. Shweta Jain (2011) Power Grid Brings GCC Economies Closer. *Gulf News*. Available at http://gulfnews.com. Accessed 25 April 2011.
30. Ottmar Edenhofer, Ramon Pichs Madruga, and Youba Sokona (2012) *Renewable Energy Sources and Climate Change Mitigation* (Cambridge: Cambridge University Press), p. 7.

Glossary

Affiliate: An entity which is directly or indirectly owned, operated, or controlled by another entity.

Air Conditioning: Cooling and dehumidifying the air in an enclosed space by use of a refrigeration unit powered by electricity or natural gas.

Alternative-Fuel Vehicle: A vehicle designed to operate on an alternative fuel (e.g. compressed natural gas, methane blend, electricity).

Appliance: A piece of equipment, commonly powered by electricity, used to perform a particular energy-driven function.

Atom: A particle of matter which cannot be broken up by chemical means. Atoms have a nucleus consisting of positively charged protons and uncharged neutrons of almost the same mass. The positive charges on the protons are balanced by a number of negatively charged electrons in motion around the nucleus.

Barrel: A unit of volume equal to 42 U.S. gallons.

Base Load: That part of electricity demand which is continuous and does not vary over a 24-hour period. Approximately equivalent to the minimum daily load.

Baseline: The reference scenario for measurable quantities from which an alternative outcome can be measured.

Biodiesel: A fuel typically made from soybean, canola, or other vegetable oils; animal fats; and recycled grease. It can serve as a substitute for petroleum-derived diesel or distillate fuel.

Bioenergy: Energy derived from any form of biomass.

Biofuels: Any liquid, gaseous, or solid fuel produced from biomass feedstocks, used primarily for transportation.

Biomass: Renewable energy from living (or recently living) plants and animals; e.g. wood chippings, crops and manure. Plants store energy from the sun while animals get their energy from the plants they eat.

Boiling Water Reactor: A common type of light water reactor, where water is allowed to boil in the core thus generating steam directly in the reactor vessel.

British Thermal Unit: The quantity of heat required to raise the temperature of 1 pound of liquid water by 1 degree Fahrenheit at which water has its greatest density.

Business as Usual: The future is projected or predicted on the assumption that operating conditions and applied policies remain what they are at present.

Capital Cost: The cost of field development and plant construction and the equipment required for industry operations.

Capital stock: Property, plant, and equipment used in the production, processing, and distribution of energy resources.

Carbon Capture and Storage: A group of technologies used to reduce CO_2 emissions from large CO_2 sources and industrial processes. Following capture, CO_2 is transported and stored in specifically selected and characterized geological formations.

Carbon Dioxide: A colourless, odourless, non-poisonous gas that is a normal part of Earth's atmosphere. Carbon dioxide is a product of fossil-fuel combustion as well as other processes. It is considered a greenhouse gas as it traps heat radiated by the Earth into the atmosphere and thereby contributes to the potential for global warming.

Carbon Footprint: The full quantity of greenhouse gases that can be attributed to an individual, a plant, a company, a product, or a whole economy.

Carbon Market: The set of organized and bilateral transactions by which countries trade credits received for greenhouse gas emission reductions.

Cells: Refers to the unencapsulated semi-conductor components of the module that convert the solar energy to electricity.

Centrifuge: A cylinder spinning at high speed to physically separate gas components of slightly different mass, e.g. uranium hexafluoride with U-235 and U-238 atoms.

Certificate: A type of permit for public convenience and necessity issued by a utility commission, which authorizes a utility or regulated company to engage in business construct facilities, provide some services, or abandon service.

Chair Reaction: A reaction that stimulates its own repetition, in particular where the neutrons originating from nuclear fission cause an ongoing series of fission reactions.

Circuit: A conductor or a system of conductors through which electric current flows.

Climate Change: A term used to refer to all forms of climatic inconsistency, but especially to significant change from one prevailing climatic condition to another. In some cases, 'climate change' has been used synonymously with the term 'global warming'.

Coal: Coal refers to a variety of solid, combustible, sedimentary, organic rocks that are composed mainly of carbon and varying amounts of other components such as hydrogen, oxygen, sulphur, and moisture.

Coal Liquefaction: A chemical process that converts coal into clean-burning liquid hydrocarbons, such as synthetic crude oil and methanol.

Coal-to-Liquids: The transformation of coal into liquid hydrocarbons.

Combined Cycle: An electric generating technology in which electricity is produced from otherwise lost waste heat exiting from one or more gas turbines.

Combined Hydroelectric Plant: A hydroelectric plant that uses both pumped water and natural stream flow for the production of power.

Commercial Sector: An energy-consuming sector that consists of service-providing facilities and equipment of businesses; government; and other private and public organization, such as religious, social, or fraternal groups. Common uses of energy associated with this sector include space heating, water heating, air conditioning, lighting, refrigeration, cooking, and running a wide variety of other equipment.

Compressed Natural Gas: Natural gas which is comprised primarily of methane, compressed to a as a fuel for natural gas pressure at or above 2,400 pounds per square inch and stored in special high-pressure containers. It is used to power vehicles.

Concentrating Photovoltaic (CPV): Technology that uses mirrors or lenses to focus sunlight onto a small area of photovoltaic cells to generate electricity.

Concentrating Solar Power: A solar energy conversion system characterized by the optical concentration of solar rays through an arrangement of mirrors to generate a high temperature working fluid.

Concession: The operating right to explore for and develop petroleum fields in consideration for a share of production.

Connection: The physical connection (e.g. transmission lines, transformers, switch gear, etc.) between two electric systems permitting the transfer of electric energy in one or both directions.

Conservation: A reduction in energy consumption that corresponds with a reduction in service demand. Service demand can include building sector end uses such as lighting, refrigeration, and heating; industrial processes; or vehicle transportation. Unlike energy efficiency, which is typically a technological measure, conservation is better associated with behaviour.

Conventional Oil and Natural Gas Production: Crude oil and natural gas that is produced by a well drilled into a geologic formation in which the reservoir and fluid characteristics permit the oil and natural gas to readily flow to the wellbore.

Conversion: Energy shows itself in numerous ways, with transformations from one type to another called energy conversions.

Core: The central part of a nuclear reactor containing the fuel elements and any moderator.

Cost: The consumption of resources such as labour time, capital, materials, fuels, etc. as the consequence of an action. In economics, all resources are valued at their opportunity cost, which is the value of the most valuable alternative use of the resources.

Crude Oil: A mixture of hydrocarbons that exists in liquid phase in natural underground reservoirs and remains liquid at atmospheric pressure after passing through surface separating facilities.

Cubic Foot: The amount of natural gas contained at standard temperature and pressure in a cube whose edges are one foot long.

Current (electric): A flow of electrons in an electrical conductor. The strength of rate of movement of the electricity is measured in amperes.

Cycle: The time period running from the startup of one reactor cycle to the start-up of the following cycle.

Dam: A physical barrier constructed across a river or waterway to control the flow of or raise the level of water. The purpose of construction may be for flood control, irrigation needs, hydroelectric power production, and/or recreation usage.

Decommissioning: Retirement of a nuclear facility, including decontamination and/or dismantlement.

Deforestation: The natural or anthropogenic process that converts forest land to non-forest.

Delivered energy: Energy generated by one system and delivered to another system through one or more transmission lines.

Deliveries (electric): Energy generated by one system and delivered to another system through one or more transmission lines.

Demand Indicator: A measure of the number of energy-consuming units, or the amount of service or output, for which energy inputs are required.

Depleted Resources: Resources that have been mined.

Deregulation: The elimination of some or all regulations from a previously regulated industry or sector of an industry.

Development: The preparation of a specific mineral deposit for commercial production.

Development costs: Costs incurred to obtain access to proved reserves and to provide facilities for extracting, treating, gathering, and storing the oil and gas.

Diesel Fuel: A fuel composed of distillates obtained in petroleum refining operation or blends of such distillates with residual oil used in motor vehicles.

Distribution: The delivery of energy to retail customers.

Distribution System: The portion of the transmission and facilities of an electric system that is dedicated to delivering electric energy to an end-user.

Drilling: The act of boring a hole to determine whether minerals are present in commercially recoverable quantities and to accomplish production of the minerals (including drilling to inject fluids).

Dry Hole: An exploratory or development well found to be incapable of producing either oil or gas in sufficient quantities to justify completion as an oil or gas well.

Dry Natural Gas: Natural gas which remains after the liquefiable hydrocarbon portion has been removed from the gas stream.

Dual Fuel Vehicle: A motor vehicle that is capable of operating on an alternative fuel and on gasoline or diesel fuel.

Duel-fired Unit: A generating unit that can produce electricity using two or more input fuels.

E85: A fuel containing a mixture of 85 per cent ethanol and 15 per cent gasoline.

Economy of Scale: The principle that larger production facilities have lower unit costs than smaller facilities.

Ecosystem: An open system of living organisms, interacting with each other and with their environment, which is capable of self-regulation to a certain degree.

Electric Generator: A facility that produces electricity, commonly expressed in kilowatthours or megawatthours. Electric generators include electric utilities and independent power producers.

Electricity: The flow of passing charge through a conductor, driven by a difference in voltage between the ends of the conductor. Electrical power is generated by work from heat in a gas or steam turbine or from wind, oceans or falling water, or produced directly from sunlight using a photovoltaic device or chemically in a fuel cell. Being a current, electricity cannot be stored and requires wires and cables for its transmission.

Electric Hybrid Vehicle: An electric vehicle that either operates solely on electricity, but contains an internal combustion motor that generates additional electricity or contains an electric system and in internal combustion system and is capable of operating on either system.

Electric Power: The rate at which electric energy is transferred. Electric power is measured by capacity and is commonly expressed in megawatts.

Electric Power Grid: A system of synchronized power providers and consumers connected by transmission and distribution lines and operated by one or more control centres.

Electricity Generation: The total amount of electricity generated by power only or combined heat and power plants.

Electricity Production: The total amount of electricity generated by a power plant.

Emissions: Anthropogenic releases of gases to the atmosphere. In the context of global climate change, they consist of important greenhouse gases (e.g. the release of carbon dioxide during fuel combustion).

End User: A firm or individual that purchases products for its own consumption and not for resale (i.e. an ultimate consumer).

Energy: The amount of work or heat delivered. Energy is classified in a variety of types and becomes available to human ends when it flows from one place to another or is converted from one type into another.

Energy Access: People are provided the ability to benefit from affordable, clean, and reliable energy services from basic human needs (cooking and heating, lighting, communication, mobility) and productive uses.

Energy Consumption: The use of energy as a source of heat or power or as a raw material input to a manufacturing process.

Energy Demand: The requirement for energy as an input to provide products and/or services.

Energy Efficiency: Something is more energy efficient if it delivers more services for the same energy input, or the same services for less energy input.

Energy Intensity: A measure of total primary energy use per unit of gross domestic product.

Energy Poverty: A lack of access to modern energy services. These services are defined as household access to electricity and clean cooking facilities.

Energy Reserves: Estimated quantities of energy sources that are demonstrated to exist with reasonable certainty on the basis of geologic and engineering data or that can reasonably be expected to exist on the basis of geologic evidence that supports projections from proved reserves.

Energy Savings: Decreasing energy intensity by changing the activities that demand energy inputs. Energy savings can be realized by technical, organizational, institutional, and structural actions and by changed behaviour.

Energy Security: The uninterrupted availability of energy sources at an affordable price with little environmental footprint.

Energy Source: Any substance or natural phenomenon that can be consumed or transformed to supply heat or power. Examples include petroleum, coal, natural gas, nuclear, biomass, electricity, wind, sunlight, geothermal, water movement, and hydrogen in fuel cells.

Enriched Uranium: Uranium in which the proportion of U-235 (to U-238) has been increased above the natural 0.7 per cent. Reactor-grade uranium is usually enriched to about 3.5 per cent U-235, weapons-grade uranium is more than 90 per cent U-235.

Enrichment: Physical process of increasing the proportion of U-235 to U-238.

Ethanol: A clear, colourless, flammable alcohol. Ethanol is typically produced biologically from biomass feed-stocks such as agricultural crops and cellulosic residues from agricultural crops or wood. It can also be produced chemically from ethylene.

Fahrenheit: A temperature scale on which the boiling point of water is at 212 degrees above zero on the scale and the freezing point is at 32 degrees above zero at standard atmospheric pressure.

Feed-in Tariff: The price per unit of electricity that a utility or power supplier has to pay for distributed or renewable electricity fed into the grid by non-utility generators.

Financing: Raising or providing money or capital by individuals, businesses, banks, venture funds, public instances, etc. for realizing a project or continuing an activity.

Fiscal Incentive: Actors (individuals, households, companies) are granted a reduction of their contribution to the public, treasury via income or other taxes.

Fissile: Capable of capturing a slow (thermal) neutron and undergoing nuclear fission.

Fissile material: Material that can be caused to undergo atomic fission when bombarded by neutrons.

Fission: A reaction when the nucleus of an atom, having captured a neutron, splits into two or more nuclei, and in so doing, releases a significant amount of energy as well as more neutrons.

Flare: A tall stack equipped with burners used as a safety device at wellheads, refining facilities, gas processing plants, and chemical plants. Flares are used for the combustion and disposal of combustible gases.

Fossil Fuel: A fuel based on carbon presumed to be originally from living matter.

Fuel: Any material substance that can be consumed to supply heat or power. Included are petroleum, coal, and natural gas (the fossil fuels), and other consumable materials, such as uranium, biomass, and hydrogen.

Gallon: A volumetric measure equal to 4 quarts used to measure fuel oil. One barrel equals 42 gallons.

Gas to Liquids: A process that combines the carbon and hydrogen elements in natural gas molecules to make synthetic liquid petroleum products, such as diesel fuel.

Gasification: A method for converting coal, petroleum biomass, wastes, or other carbon-containing materials into a gas that can be burned to generate power or processed into chemicals and fuels.

Geothermal: Energy available as heat emitted from within the earth's crust, usually in the form of hot water or stream.

Giga: One billion.

Gigawatt: One billion watts or one thousand megawatts.

Gigawatt-electric: One billion watts of electric capacity.

Gigawatthour: One billion watthours.

Global Warming: An increase in the near surface temperature of the Earth. Global warming has occurred in the distant past as the result of natural influences but the term is today most often used to refer to the warming some scientists predict will occur as a result of increased anthropogenic emissions of greenhouse gases.

Governance: Governance is a comprehensive and inclusive concept of the full range of means for deciding, managing, and implementing polices and measures. It recognizes the contributions of various levels of government (global, regional, and local) and the contributing roles of the private sector, of non-governmental actors, and of civil society to addressing the many types of issues facing the global community.

Green Energy Purchase: Voluntary purchase of renewable energy, usually electricity, by residential, commercial, government, or industrial consumers, either directly from an energy trader or utility company, from a third-party renewable energy generator, or indirectly via trading of renewable energy certificates.

Greenhouse Effect: The result of water vapour, carbon dioxide, and other atmospheric gases trapping radiant energy, thereby keeping the Earth's surface warmer than it would otherwise be.

Greenhouse gases: Those gases, such as water vapour, carbon dioxide, nitrous oxide, methane, and sulphur hexafluoride, that are transparent to solar radiation but opaque to long-wave radiation, thus preventing long-wave radiant energy from leaving Earth's atmosphere. The net effect is a trapping of absorbed radiation and a tendency to warm the planet's surface.

Grid: The layout of an electrical distribution system. A network consisting of wires, switches, and transformers to transmit electricity from power sources to power users.

Heavy Water: Water containing a significantly greater proportion of heavy hydrogen atoms to ordinary hydrogen atoms than is found in ordinary water. Heavy water is used as a moderator in some reactors because it slows neutrons effectively and also has a low cross-section for absorption of neutrons.

High-enriched Uranium: Uranium enriched to 20 per cent U-235 or more.

Household Energy Expenditures: The total amount of funds spent for energy consumed in, or delivered to, a housing unit during a given period of time.

Hydrocarbon: An organic chemical compound of hydrogen and carbon in the gaseous, liquid, or solid phase.

Hydroelectric Power: The energy of water moving from higher to lower elevations that is converted into mechanical energy through a turbine or other device that is either used directly for mechanical work or more commonly to operate a generator that produces electricity.

Idle Capacity: The component of operable capacity that is not in operation and not under active repair, but capable of being placed in operation within short period of time.

Improved Recovery: Extraction of crude oil or natural gas by any method other than those that rely primarily on natural reservoir pressure, gas lift, or a system of pumps.

Industrial Sector: An energy-consuming sector that consists of all facilities and equipment used for producing, processing, or assembling goods.

Interconnected System: Two or more electric systems having a common transmission line that permits a flow of energy between them. The physical connection of the electric power transmission facilities allows for the sale or exchange of energy.

Investment Tax Credit: A taxation measure that allows investments in renewable energy to be fully or partially deducted from the tax obligations or income of a project developer, industry, building owner, etc.

Isotope: An atomic form of an element having a particular number of neutrons. Different isotopes of an element have the same number of protons but different numbers of neutrons and hence different atomic masses.

Kilowatt: One thousand watts.

Kyoto Protocol: The result of negotiations at the third Conference of the Parties in Kyoto held in Japan (December 1997). It sets binding greenhouse gas emissions targets for countries that sign and ratify the agreement.

Life Extension: Restoration or refurbishment of a plant to its original performance without the installation of new combustion technologies.

Light Water: Ordinary water as distinct from heavy water.

Light Water Reactor: A nuclear reactor that uses water as the primary coolant and moderator, with slightly enriched uranium as fuel.

Lignite: The lowest rank of coal, often referred to as brown coal, used almost exclusively as fuel for stream-electric power generation.

Line Loss: Electric energy lost because of the transmission of electricity.

Liquefied Natural Gas: Natural gas that has been liquefied by reducing its temperature at atmospheric pressure. In this way, the space requirements for storage and transport are reduced by a factor of over 600.

Liquid Fuels: All petroleum products, natural gas liquids, bio-fuels, and liquids derived from other hydrocarbon sources.

Load: The demand for electricity by (thousands to millions) power users at the same moment aggregated and raised by the losses in transport and delivery, and to be supplied by the integrated power supply system.

Low-Carbon Technologies: Technologies that produce low – or zero – greenhouse gas emissions while operating.

Manufacturing: An energy-consuming subsector of the industrial sector that consists of all facilities and equipment engaged in the mechanical, physical, chemical, or electronic transformation of materials, substances, or components into new products.

Megawatt: One million watts of electricity.

Methane: A colourless, flammable, odourless hydrocarbon gas which is the major component of natural gas. It is also an important source of hydrogen in various industrial processes.

Mineral: Any of the various naturally occurring in organic substances, such as metals, salt, and stone, sulphur, and water, usually obtained from the Earth.

Mineral rights: The ownership of the minerals beneath the Earth's surface with the right to remove them.

Mining: An energy-consuming sub-sector of the industrial sector that consists of all facilities and equipment used to extract energy and mineral resources.

Mitigation: Technological change and changes in activities that reduce resource inputs and emissions per unit of output.

Natural Decline Rate: The base production decline rate of an oil or gas field without intervention to enhance production.

Natural Gas: A gaseous mixture of hydrocarbon compounds, the primary one being methane.

Natural Gas Liquids: The liquid or liquefied hydrocarbons produced in the manufacture, purification, and stabilization of natural gas.

Natural Gas Used for Injection: Natural gas used to pressurize crude oil reservoirs in an attempt to increase oil recovery or in instances where there is no market for the natural gas. Natural gas used for injection is sometimes referred to as re-pressuring.

Natural Uranium: Uranium with the U-235 isotope present at a concentration of 0.711 per cent (by weight), that is, uranium with its isotopic content exactly as it is found in nature.

Nominal Price: The price paid for a product or service at the time of the transaction. Nominal prices are those that have not been adjusted to remove the effect of changes in the purchasing power of the dollar, they reflect buying power in the year in which the transaction occurred.

Nuclear Electric Power (Nuclear Power): Electricity generated by the use of the thermal energy released from the fission of nuclear fuel in a reactor.

Nuclear Fuel: Fissionable materials that have been enriched to such a composition that, when placed in a nuclear reactor, will support a self-sustaining fission chain reaction, producing heat in a controlled manner for process use.

Nuclear Reactor: An apparatus in which a nuclear fission chain reaction can be initiated, controlled, and sustained at a specific rate. A reactor includes fuel (fissionable material), moderating material to control the rate of fission, a heavy-walled pressure vessel to house reactor components, shielding to protect personnel, a system to conduct heat away from the reactor, and instrumentation for monitoring and controlling the reactor's systems.

Ocean Energy Systems: Energy conversion technologies that harness the energy in tides, waves, and thermal gradients in the oceans.

Offshore: The geographic area that lies seaward of the coastline. In general, the coastline is the line of ordinary low water along with that portion of the coast that is in direct contact with the open sea or the line marking the seaward limit of inland water.

Oil: A mixture of hydrocarbons usually existing in the liquid state in natural underground pools or reservoirs. Gas is often found in association with oil.

Oil Reservoir: An underground pool of liquid consisting of hydrocarbons, sulphur, oxygen, and nitrogen trapped within a geological formation and protected from evaporation by the overlying mineral strata.

Oil Shale: A sedimentary rock containing kerogen, a solid organic material.

Oil Well: A well completed for the production of crude oil from at least one oil zone or reservoir.

Parent Company: An affiliated company that exercises ultimate control over a business entity, either directly or indirectly, through one or more intermediaries.

Payback: Mostly used in investment appraisal as financial payback, which is the time needed to repay the initial investment by the returns of a project.

Peak Demand: The maximum load during a specified period of time.

Petrochemical feed-stocks: Chemical feed-stocks derived from petroleum principally for the manufacture of chemicals, synthetic rubber, and a variety of plastics.

Petrochemicals: Organic and in organic compounds and mixtures that include but are not limited to organic chemicals, cyclic intermediates, plastics and resins, synthetic fibres, organic dyes, organic pigments, detergents, surface active agents, carbon black, and ammonia.

Petroleum: A broadly defined class of liquid hydrocarbon mixtures. This includes crude oil, lease condensate, unfinished oils, refined products obtained from the processing of crude oil and natural gas plant liquids.

Petroleum Products: Petroleum products are obtained from the processing of crude oil, natural gas, and other hydrocarbon compounds.

Petroleum Refinery: An installation that manufactures finished petroleum products from crude oil, unfinished oils, natural gas liquids, other hydrocarbons, and alcohol.

Photovoltaic and Solar Thermal Energy: Energy radiated by the sun as electromagnetic waves that is converted at electric utilities into electricity by means of solar cells or concentrating collectors.

Photovoltaic Cell: An electronic device consisting of layers of semiconductor materials fabricated to form a junction (adjacent layers of materials with different electronic characteristics) and electrical contacts and being capable of converting incident light directly into electricity.

Pipeline (Natural Gas): A continuous pipe conduit, complete with such equipment as valves, compressor stations, communications systems, and metres for transporting natural and/or supplemental gas from one point to another.

Pipeline (Petroleum): Crude oil and product pipelines used to transport crude oil and petroleum products.

Plutonium: A heavy, fissionable, radioactive, metallic element that occurs naturally in trace amounts. It can also result as a by-product of the fission reaction in a uranium-fuel nuclear reactor and can be recovered for future use.

Power: The rate of producing, transferring, or using energy, most commonly associated with electricity. Power is measured in watts and often expressed in kilowatts or megawatts.

Price: The amount of money or consideration-in-kind for which a service is bought, sold, or offered for sale.

Primary Energy: Energy in the form that it is first accounted for in a statistical energy balance, before any transformation to secondary or tertiary forms of energy.

Processing: Uranium recovery operations whether at a mill, an in situ leach, by-product plant, or other type of recovery operation.

Production Capacity: The amount of product that can be produced from processing facilities.

Production Costs: Costs incurred to operate and maintain wells and related equipment and facilities, including depreciation and applicable operating costs of support equipment and facilities and other costs of operating and maintaining those wells and related equipment and facilities.

Proven Energy Reserves: Estimated quantities of energy sources that analysis of geologic and engineering data demonstrates with reasonable certainty are recoverable under existing economic and operating conditions.

Public Good: Public goods are simultaneously used by several parties (opposite to private goods). Some public goods are fully free from rivalry in use; for others the use by some subtract from the availability for others, creating congestion. Access to public goods may be restricted dependent on whether public goods are commons, state-owned or no one's case. The atmosphere and climate are the ultimate public goods of mankind. Many renewable energy sources are also public goods.

Public–Private Partnerships: Arrangements typified by joint working between the public and private sectors. In the broadest sense, they cover all types of collaboration across the interface between the public and private sectors to deliver services or infrastructure.

Quota: Established quotas obligate designated parties (generators or suppliers) to meet minimum (often gradually increasing) renewable energy targets, generally expressed as percentages of total supplies or as an amount of renewable energy capacity, with costs borne by consumers.

Radiation: The transfer of heat through matter or space by means of electromagnetic waves.

Radioactivity: The spontaneous decay of an unstable atomic nucleus, giving rise to the emission of radiation.

Radioactive waste: Materials left over from making nuclear energy. Radioactive waste can destroy living organisms if it is not stored safely.

Real Price: A price that has been adjusted to remove the effect of changes in the purchasing power of the dollar. Real prices, which are expressed in constant dollars, usually reflect buying power relative to a base year.

Recycling: The process of converting materials that are no longer useful as designed or intended into a new product.

Reference Month/Year: The calendar month/year to which the reported cost, price, and volume information relates.

Refinery: An installation that manufactures finished petroleum products from crude oil, unfinished oils, natural gas liquids, other hydrocarbons, and oxygenates.

Regulation: The governmental function of controlling or directing economic entities through the process of rulemaking and adjudication.

Re-injected: The forcing of gas under pressure into an oil reservoir in an attempt to increase recovery.

Reliability: The degree of performance according to imposed standards or expectations.

Renewable Energy: Energy that is derived from natural processes (e.g. sunlight and wind) that are replenished at a higher rate than they are consumed. Solar, wind, geothermal, hydro, and biomass are common sources of renewable energy.

Renewable Energy Target: An official commitment, plan, or goal set by a government (at local, state, national, or regional level) to achieve a certain amount of renewable energy by a future date. Some targets are legislated while others are set by regulatory agencies or ministries.

Renewable Portfolio Standard: A measure requiring that a minimum percentage of total electricity or heat sold, or generation capacity installed, be provided using renewable energy sources. Obligated utilities are required to ensure that the target is met, if it is not, a fine is usually levied.

Reprocessing: Chemical treatment of used reactor fuel to separate uranium and plutonium from the small quantity of fission products, leaving a much reduced quantity of high-level waste.

Research and Development: Basic and applied research in the sciences and engineering and the design and development of prototypes and processes.

Reservoir: A porous and permeable underground formation containing an individual and separate natural accumulation of producible hydrocarbons (crude oil and/or natural gas), which is confined by impermeable rock or water barriers and is characterized by a single natural pressure system.

Residential Sector: An energy-consuming sector that consists of living quarters for private households. Common uses of energy associated with this sector include space heating, water heating, air conditioning, lighting, refrigeration, cooking, and running a variety of other appliances.

Retailer: A firm that carries on the trade or business of purchasing refined petroleum products and reselling them to ultimate consumers without substantially changing their form.

Revenue: The total amount of money received by an entity from sales of its products and/or services; gains from the sales or exchanges of assets, interest, and dividends earned on investments; and other increases in the owner's equity.

Scenario: A plausible description of how the future may develop based on a coherent and internally consistent set of assumptions about key relationships and driving forces (e.g. rate of technological change, prices) on social and economic development, energy use, etc. Scenarios are neither predictions nor forecasts, but are useful to provide a view of the implications of alternative developments and actions.

Shale Gas: Natural gas produced from wells that are open to shale formations. The shale acts as both the source and the reservoir for the natural gas.

Smart Grid: An electricity network that uses digital and other advanced technologies to monitor and manage the transport of electricity from all generation sources to meet the varying electricity demands of end users.

Solar Energy: The radiant energy of the sun, which can be converted into other forms of energy, such as heat or electricity.

Solar Power Tower: A solar energy conversion system that uses a large field of independently adjustable mirror to focus solar rays on a near singly point atop a fixed tower (receiver).

Solar Radiation: A general term for the visible and near visible electromagnetic radiation that is emitted by the sun.

Solar Thermal Collector: A device designed to receive solar radiation and convert it to thermal energy.

Solar Thermal Panels: A system that actively concentrates thermal energy from the sun by means of solar collector panels. The panels typically consist of fat, sun-oriented boxes with transparent covers, containing water tubes of air baffles under a blackened heat absorbent panel.

Solar Water Heaters: Solar collectors, usually rooftop mounted but also on-ground at a larger scale, that heat water and store it in a tank for later use as hot water or for circulation to provide space or process heating.

Spent Fuel: Irradiated fuel that is permanently discharged from a reactor. Except for possible reprocessing, this fuel must eventually be removed from its temporary storage location at the reactor site and placed in a permanent repository.

Standards: Set of rules or codes mandating or defining product performance (e.g. grades, dimensions, characteristics, test methods, and rules for use).

Subsidiary: An entity directly or indirectly controlled by a parent company which owns 50 per cent or more of its voting stock.

Subsidies: Any government action directed primarily at the energy sector that lowers the cost of energy production, raises the price received by energy producers or lowers the price paid by energy consumers.

Sulphur: A yellowish non-metallic element. It is present at various levels of concentration in many fossil fuels whose combustion releases sulphur compounds that are considered harmful to the environment.

Tar Sands: Naturally occurring bitumen-impregnated sands that yield mixtures of liquid hydrocarbon and that require further processing other than mechanical blending before becoming finished petroleum products.

Tariff: A published volume of rate schedules and general terms and conditions under which a product or service will be supplied.

Technological Change: Mostly considered as technological improvement, that is, more or better goods and services can be provided from a given amount of resources.

Technology: The practical application of knowledge to achieve particular tasks that employs technical hardware, equipment, and information.

Technology Transfer: The exchange of knowledge, hardware, and associated software, money and goods among stakeholders, which leads to the spread of technology for adaptation or mitigation. The term encompasses both diffusion of technologies and technological cooperation across and within countries.

Thermal: A term used to identify a type of electric generating station, capacity, capability, or output in which the source of energy for the prime mover is heat.

Transmission: An interconnected group of lines and associated equipment for the movement or transfer of electric energy between points of supply and points at which it is transformed for delivery to customers or is delivered to other electric systems.

Transmission and Distribution Loss: Electric energy lost due to the transmission and distribution of electricity.

Transportation Sector: An energy-consuming sector that consists of all vehicles whose primary purpose is transporting people and/or goods from one physical location to another.

Uranium: A heavy, naturally radioactive, metallic element. Its two principally occurring isotopes are uranium-235 and uranium-238. Uranium-235 is indispensable to the nuclear industry because it is the only isotope existing in nature, to any appreciable extent, that is fissionable by thermal neutrons. Uranium-238 is also important because it absorbs neutrons to produce a radioactive isotope that subsequently decays to the isotope plutonium-239, which also is fissionable by thermal neutrons.

Uranium Mill: A plant where uranium is separated from ore taken from mines.

Uranium Ore: Rock containing uranium mineralization in concentrations that can be mined economically.

Used Fuel: Fuel assemblies removed from a reactor after several years' use.

Vent: An opening at the surface of the Earth through which materials and energy flow.

Voltage: The difference in electrical potential between any two conductors or between a conductor and ground. It is a measure of the electric energy per electron that electrons can acquire and/or give up as they move between the two conductors.

Watt: The unit of electrical power equal to one ampere under a pressure of one volt. A Watt is equal to 1/746 horse power.

Watthour: The electrical energy unit of measure equal to one watt of power supplied to, or taken from, an electric circuit steadily for one hour.

Wind Energy: Kinetic energy present in wind motion that can be converted to mechanical energy for driving pumps, mills, and electric power generators.

Wind Turbine: Wind energy conversion device that produces electricity; typically three blades rotating about a horizontal axis and positioned up-wind of the supporting tower.

Wind Farm, Wind Project, Wind Power Plant: A group of wind turbines interconnected to a common utility system through a system of transformers, distribution lines, and usually one substation.

Yellowcake: A natural uranium concentrate that takes its name from its colour and texture.

Index

Printed and bound by CPI Group (UK) Ltd, Croydon, CR0 4YY